W9-AJU-036

GENETICS DEMYSTIFIED

JAN - 2006

Demystified Series

GENETICS DEMYSTIFIED

EDWARD WILLETT

McGRAW-HILL
New York   Chicago   San Francisco   Lisbon   London
Madrid   Mexico City   Milan   New Delhi   San Juan
Seoul   Singapore   Sydney   Toronto

The McGraw-Hill Companies

Cataloging-in-Publication Data is on file with the Library of Congress

1 2 3 4 5 6 7 8 9 0   DOC/DOC   0 1 0 9 8 7 6 5

ISBN 0-07-145930-8

The sponsoring editor for this book was Judy Bass and the production supervisor was Pamela A. Pelton. It was set in Times Roman by Fine Composition. The art director for the cover was Margaret Webster-Shapiro; the cover designer was Handel Low.

Printed and bound by RR Donnelley.

 This book is printed on recycled, acid-free paper containing a minimum of 50% recycled, de-inked fiber.

McGraw-Hill books are available at special quantity discounts to use as premiums and sales promotions, or for use in corporate training programs. For more information, please write to the Director of Special Sales, McGraw-Hill Professional, Two Penn Plaza, New York, NY 10121-2298. Or contact your local bookstore.

*To my late grandparents, Roy Edward Spears, Laura Edwin Umstattd,
Ewan Chambers Willett, and Bessie Brown—thanks for the genes!*

CONTENTS

PREFACE

As a weekly science columnist, I try to stay at least peripherally aware of research news from all over the world. It's constantly fascinating and intriguing, but also a little frustrating, because science is advancing so quickly on so many fronts that more than once scientific beliefs I have confidently restated in writing have been overthrown, or at least cast into doubt, by new research—sometimes within weeks, once or twice almost before the newspaper ink dried.

Genetics is one of those fields advancing at a furious pace. If some of its advances have recently cast into doubt some of the theories I've treated in this book as well-established, I hope the reader will understand—and take that fact as good reason to continue to study and follow this exciting, ever-changing (and world-changing) area of scientific endeavor.

Learning new things is what I enjoy most about being a science writer. I hope it's also what you enjoy most about reading this book.

EDWARD WILLETT

GENETICS DEMYSTIFIED

1

Mendelism and Classical Genetics

The concept of the gene came largely from the work of one man—an Augustinian monk named Gregor Mendel.

Before Mendel conducted his groundbreaking experiments on pea plants in the 1860s, everyone knew that offspring tended to inherit some characteristics of their parents. You could hardly fail to notice something like that, since everyone is, after all, an offspring, and many people are parents. But exactly how it happened was unknown.

Heredity Happens, but How?

This lack of knowledge about the mechanism of heredity hampered other areas of science. For instance, in his 1859 book *On the Origin of Species*, Charles Darwin claimed that species of living things slowly changed—evolved—over

time. He said this happened because occasionally members of a species were born a little bit different than their fellows. (For example, some individuals within a species of grazing animal might be born with slightly longer necks than other individuals.) If that difference had survival benefits—for example, a longer neck to help the animals reach more leaves to feed on—the slightly different individuals were more likely to pass on their difference to their offspring. The offspring would also be more likely to survive, until, over time, that slight difference was present in almost all members of that species. (In this example, they would all have longer necks than their distant ancestors). This process is called *natural selection*.

Some scientists objected to Darwin's theory on the basis that he could not explain exactly how offspring inherited characteristics from their parents. Darwin admitted he could not, and that the question puzzled him.

But even while arguments raged around Darwin's theory, Mendel was laying the groundwork for a whole new field of science that would shed light on every aspect of biology, including evolution.

The Scientific Monk

Gregor Mendel (see Fig. 1-1) was born Johann Mendel on July 22, 1822 in the village of Heinzendorft, in the region of the Austria-Hungary Empire called Moravia. (Today it's known as Hyncice and is in the Czech Republic.) He came from a family of peasant farmers, but his parish priest and the local teacher noted his intelligence, and helped him get secondary schooling, unusual for someone of his status. After his father was injured in an accident in 1838, however, there was no more money for school. He managed to earn enough money on his own to start college courses, but both his work and his health suffered because of the constant struggle for funds.

One of his professors suggested that he join the Augustinians, whose main work was teaching, because the order would pay for his schooling. Mendel joined the Augustinian monastery in the town of Brünn (now Brno, Czech Republic), the Abbey of St. Thomas, in 1843 at the age of 21. As required by the order, he took a new name, Gregor.

The Abbey was a remarkable place. The friars there had access to scientific instruments, an excellent botanical collection, and an extensive library. The abbot, C. F. Napp, was president of the Pomological and Aenological Association, and

Fig. 1-1.   Gregor Mendel.

a close associate of F. Diebl, professor of agriculture at the University of Brünn. Abbot Napp shared a love of plants with Mendel, and it was thanks to him that Mendel was able to study at the University of Vienna from 1851 to 1853. Mendel attended courses on plant physiology and experimental physics, among other topics. His professors emphasized the importance of studying nature through experiments underpinned by mathematical models.

Although the Abbey had sent Mendel to Vienna partly so that he could pass his teaching examination (which he'd failed in 1850), he failed it again upon his return. That limited him to substitute teaching. On the bright side, it also freed up more time for what he really loved: gardening.

Abbot Napp allowed Mendel to make use of a portion of the Abbey's large garden and greenhouse in any way he liked. What Mendel did with those resources was apply the rigorous scientific methods of physics to the problem of heredity.

Mendel's Experiments

Like other plant breeders, Mendel knew that when you crossed plants with different characteristics, the resultant hybrids sometimes showed the characteristics of both parents—but not always. Sometimes traits from one parent would seem to disappear, only to reappear in later generations. Mendel wondered if there was a pattern to this phenomenon and decided to find out.

For his experiments, he chose common garden peas, *Pisum sativum*, because they had large flowers which made them easier to work with, and a wide range of variations he could map. As well, *Pisum sativum* is self-fertile, and breeds true: that is, an individual plant's offspring will closely resemble it unless the plant is artificially fertilized with pollen from another plant.

Mendel decided to focus on seven characteristics that he thought stood out "clearly and definitely" in the plants:

1. Form of the ripe seeds
    a. round or roundish, or
    b. angular and wrinkled
2. Color of the seed
    a. pale yellow, bright yellow, and orange, or
    b. green
3. Color of the seed coat
    a. white or
    b. gray, grey brown, or leather brown, with or without violet spotting
4. Form of the ripe pods
    a. simply inflated or
    b. deeply constricted and more or less wrinkled
5. Color of the unripe pod
    a. light to dark green or
    b. vividly yellow
6. Position of the flowers
    a. axial (distributed along the main stem) or
    b. terminal (bunched at the top of the stem)
7. Length of the stem
    a. long (six to seven feet) or
    b. short (¾ to one foot)

Over eight years, Mendel grew and studied almost 30,000 pea plants, following some plants through as many as seven generations. He mated plants that differed in particular characteristics (extremely painstaking work that involved transferring pollen from one plant to another), then counted the number of offspring that showed each form of the characteristic he was testing. Then he let the hybrids and their offspring fertilize themselves.

Finally, he used mathematics to discover general rules about the way the various characteristics were inherited.

Emerging Patterns

As Mendel analyzed his data, patterns emerged. Crossing tall plants with short ones, for example, always produced tall plants. If the hybrid tall plants were allowed to self-fertilize, however, the next generation had about one short plant in every four. In the next generation after that—and in more generations after that—the short plants always produced more short plants, one-third of the tall plants produced only tall plants, and the remaining two-thirds of the plants produced both tall and short plants, in that same ratio of three to one (see Fig. 1-2).

Mendel got those results with every one of the seven traits he chose to study. From this he concluded:

1. Characters or traits from the parent pea plants passed as unmodified units or "factors" to successive generations in set ratios.
2. Each individual plant contained two factors that specified the form of each trait. One factor came from the egg and one from the sperm.
3. Since each parent plant would also have two genes, the parents' pairs of genes had to separate during the forming of the sex cells so that each sex cell only contained one form of the gene. This is now known as the *principle of segregation.*
4. Chance determined which of the four possible combinations of factors each offspring received.

Mendel's statistical analysis showed that one form of each trait was about three times more common in all generations other than the first one. Mendel called the trait that appeared more often *dominant*, and the one that appeared less often *recessive*. When both were present in an individual, the plant would take on the characteristic of the dominant factor. However, although the dominant factor masked the recessive one in that case, it didn't alter it in any way. That meant the trait associated with the recessive factor could still show up in a later generation, when chance produced a plant with two copies of it.

Mendel also discovered that each trait he studied was independent of the others. That is, whether a plant was tall or short had no bearing on whether it

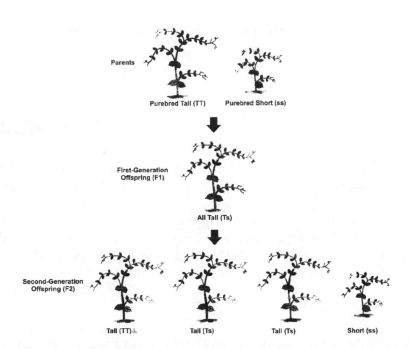

Parents

Purebred Tall (TT)   Purebred Short (ss)

First-Generation
Offspring (F1)

All Tall (Ts)

Second-Generation
Offspring (F2)

Tall (TT)   Tall (Ts)   Tall (Ts)   Short (ss)

Fig. 1-2. Gregor Mendel crossed pairs of pea plants with different characteristics
and charted the results. In this example, tall plants that always produced
tall offspring (TT) were mated with short plants that always produced
short offspring (ss). In the first generation (F1) all the plants were tall,
but Mendel came to realize they still contained the shortness "factor"—
what we now call a gene: the tallness gene simply masked it. When those
plants (Ts) were cross-fertilized, they produced a mixture of tall and
short plants in a ratio of 3:1—three tall (TT, Ts, Ts) and one short (ss).

produced round or wrinkled seeds. This is now known as the *principle of independent assortment*.

Mendel wrote a paper describing his work in 1865. Called "Experiments in
Plant Hybridization," it was published in the *Journal of the Brünn Society of
Natural Science*. It went essentially unnoticed until 1900, when three researchers, Hugo de Vries, Erich Tschermak von Seysenegg, and Karl Correns independently rediscovered Mendel's work and laws.

In the years since, we've applied new terminology to Mendel's discoveries.
His factors are now called *genes*. Each possible form a gene can take is called
an *allele*. Organisms that contain two copies of the allele are called *homozygous*.
Organisms that contain copies of two different alleles are called *heterozygous*.

The Punnett Square

One method of determining all possible combinations of alleles is through the use of a *Punnett Square*, invented by Reginald C. Punnett, who was both a mathematician and a biologist. The Punnett Square uses a table format that lists all of the possible alleles from one parent along the top, and all of the possible alleles from the other parent down the left side. The squares then show all the combinations.

For example, here is the Punnett Square for a self-fertilized heterozygous pea plant like the first-generation one in Fig. 1-2.

	T	s
T	TT (tall plant)	Ts (tall plant)
s	Ts (tall plant)	ss (short plant)

Once again, we see the 3:1 ratio of the dominant trait over the recessive trait.

A cross between a homozygous short plant (ss) and a heterozygous tall plant (Ts) produces this Punnett Square, with the offspring split evenly between tall and short.

	T	s
T	Ts (tall plant)	Ts (tall plant)
s	ss (short plant)	ss (short plant)

On the other hand, a cross between a homozygous tall plant (TT) and a heterozygous tall plant (Ts) produces only tall plants.

	T	T
T	TT (tall plant)	TT (tall plant)
s	Ts (tall plant)	TT (tall plant)

The Punnett Square is a simple but powerful tool for calculating the expected statistical distribution of particular traits—and it works with traits where there are more than just two alleles involved, too.

Extensions to Mendel's Laws

Mendel's discoveries were the starting point for the science of genetics. (The term genetics was coined in 1906, by British zoologist William Bateson, who read Mendel's paper in 1900, recognized its importance, and helped bring Mendel's work to the attention of the scientific community. Bateson defined genetics as "the elucidation of the phenomena of heredity and variation." Interestingly, the term gene wasn't coined until 1909, when Danish biologist Wilhelm Johansson proposed that it be used instead of Mendel's vague word "factor.")

Mendel's work provided a solid foundation for the many discoveries that came after it—discoveries which are the focus of the rest of this book. Before we move on, though, a couple of important exceptions to his simple rules of inheritance should be mentioned.

Mendel was fortunate in that the traits he studied in pea plants are examples of complete dominance: when a dominant allele was present in an individual plant, that allele completely masked the presence of a recessive allele. That's not always the case. In some species, certain traits are controlled by genes which exhibit incomplete dominance. One example is the flower color in primroses. Primroses with red flowers have two copies of the dominant "red" allele, while primroses with white flowers have two copies of the recessive "white" allele. Primroses with a copy of each allele, however, don't have red flowers: they have pink flowers. The red allele does not result in the production of enough red pigment to completely color the flowers.

There are also traits that are codominant: in heterozygous individuals, both alleles are expressed. That's why there are people with A-type blood, B-type blood, and AB-type blood. Those with AB-type blood have blood with characteristics of both type A and type B.

Blood type is also an example of a multiple-allele series. Instead of there being just two alleles, A and B, there is also a third one, O. However, each individual still inherits only two alleles. O is rare because an individual has to have two O alleles in order to have O-type blood.

Even more alleles control some traits. In fact, it now appears that multiple-allele traits are more common than two-allele traits.

Despite these exceptions, Mendel's work remains a classic example of solid experimental work. Very few scientists can say they gave birth to a whole new scientific discipline.

Mendel, alas, didn't live to see it. He died in 1884, just about the time that scientists were beginning to better understand the basic unit of life in which heredity is expressed: the cell.

In the next chapter, we'll examine this basic unit of life in more detail.

Quiz

1. The basic mechanism of Darwinian evolution is called
   (a) acquired heredity.
   (b) natural selection.
   (c) trait inflation.
   (d) mate mixing.

2. What type of plants did Mendel use in his experiments?
   (a) corn
   (b) petunias
   (c) garden peas
   (d) strawberries

3. How many traits did Mendel study?
   (a) four
   (b) five
   (c) six
   (d) seven

4. In what ratio did the plants with the dominant trait appear in the second offspring generation?
   (a) 5:1
   (b) 3:1
   (c) 2:1
   (d) 1:1

5. Mendel discovered that the two "factors" controlling each of the traits he studied were split apart during the formation of sex cells, so that each parent plant gave only one of its factors to the resulting offspring. What is this principle called?
   (a) The principle of sexual selection.
   (b) The law of unintended consequences.
   (c) The principle of segregation.
   (d) The uncertainty principle.

6. Mendel also discovered that each trait he studied was independent of the others. That is, whether a plant was tall or short had no bearing on whether it produced round or wrinkled seeds. What is this principle called?
   (a) The principle of independent assortment.
   (b) The principle of segregation.
   (c) The principle of integral heredity.
   (d) The principle of genetic conservation.

7. An individual with two different forms of a particular gene is called
   (a) heterosexual.
   (b) heterodyne.
   (c) heteroptical.
   (d) heterozygous.

8. When a dominant trait does not completely mask a recessive trait, this is known as
   (a) incomplete recessiveness.
   (b) incomplete dominance.
   (c) weak expression.
   (d) mixed messaging.

9. When two traits are equally expressed in an individual, they are said to be
   (a) competing.
   (b) divisive.
   (c) recidivist.
   (b) codominant.

10. An allele is
    (a) a type of garden flower.
    (b) an Augustinian prayer.
    (c) one form of a gene.
    (d) a plant's female sexual organ.

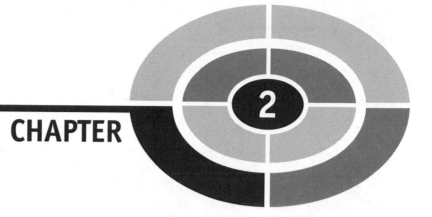

The Cell— The Basic Unit of Life

There was another advance in biological science in the 19th century that pre-dated Mendel's experiments, but would prove just as crucial in unraveling the mystery of heredity: the realization that the basic unit of life was the cell, just as atoms are the basic building blocks of matter.

Cell Theory

The development of *cell theory* grew out of the invention of the microscope at the beginning of the 17th century. In 1665, the English physicist and microscopist Robert Hooke published the book *Micrographia,* devoted to what he had observed through his microscope. Among other things, he described the structure of a slice of cork, which his microscope had revealed to be an array of tiny compartments like little rooms. Hooke called these tiny compartments "cells," from the Latin *cella* for "a small room."

Although he provided the word, Hooke didn't have a clue about the true nature of cells: He thought the cell walls he saw (which were actually thickened and dead) were passages through which fluids flowed.

Animalcules and Nuclei

Probably the first person to observe living cells, also in the late 1600s, was Dutch shopkeeper Anton van Leeuwenhoek, whose hobby was building and using microscopes. Although by modern standards they were pretty primitive, his microscopes had excellent lenses through which he observed single-cell organisms he called "animalcules."

As microscopes improved, so did observations of the cell. Although cell *nuclei* were probably observed as early as the beginning of the 18th century, it was Scottish botanist Robert Brown who realized, in 1831, that the opaque spot he'd been seeing in orchid cells was an essential part of all living cells. Cell nuclei would eventually become central (sorry) to the study of genetics.

Schleiden and Schwann

In 1838, German botanist Matthias Jakob Schleiden suggested that every part of a plant is made up of cells or their products. The following year zoologist Theodor Schwann reached the same conclusion regarding animals, stating, "there is one universal principle of development for the elementary parts of organisms…and this principle is in the formation of cells."

Schleiden and Schwann are now considered the fathers of cell theory, but another important piece of the theory was provided by German pathophysiologist Rudolph Virchow, who formulated the Latin aphorism *omnis cellula e cellula,* meaning "every cell from a pre-existing cell." In 1858 Virchow wrote, "Every

The Three Tenets of Cell Theory

1. All living things are made of cells.
2. Cells only arise from pre-existing cells by division. (In other words, life doesn't arise spontaneously out of nonliving things.)
3. Cells are made of similar compounds with similar characteristics and biochemistries.

animal appears as a sum of vital units each of which bears in itself the complete characteristics of life."

Parts of the Cell

Biologists continued to examine cells with better and better microscopic equipment and, just as importantly, better techniques for staining and preparing specimens for microscopic viewing. They began to identify more and more structures inside living cells. They also discovered an important difference between the cells of multicellular organisms (e.g., plants, animals) and the simplest single-cell organisms (e.g., bacteria).

The cells of multicellular organisms, and some single-celled organisms like fungi and protozoa, contain a well-defined cell nucleus bounded by a membrane, and a number of smaller structures called *organelles,* also bounded by membranes. Cells with these structures are called *eukaryotic cells.* Bacteria have only a single membrane (the outer cell wall), and lack both a defined nucleus and organelles. These kinds of cells are called *prokaryotic.*

CROSS-REFERENCE

Most of this book deals with the genetics of eukaryotic cells, but Chapter 9, "Bacteria—A Different Way of Doing Things," is devoted to prokaryotic genetics.

By the end of the 19th century, all of the principal organelles had been identified. Table 2-1 lists these organelles and describes their function. (Note that not every type of cell has every type of organelle.) Fig. 2-1 illustrates their location and appearance in a typical cell.

Table 2-1. Cell Organelles

Organelle Name	Description
Cell wall	Found mainly in plants, the cell wall maintains the shape of the cell. In plants, it also serves as the organism's skeleton—it's what makes plants strong enough to stand upright. There's typically a primary cell wall that's very elastic, and a secondary cell wall that forms around the primary one once the cell has stopped growing. Cell walls often survive intact long after the cells that formed them are gone. It was these dead cell walls that Robert Hooke observed in sliced cork.
Plasma membrane	This outer membrane, common to all cells (it's inside the cell wall in cells that have cell walls), separates the cell from other cells and the environment. It helps control the passage of substances into and out of the cell's interior.
Cytoplasm	Originally called "protoplasm," cytoplasm is a collective term for all the stuff inside the plasma membrane, including the other organelles listed below. Everything is suspended in *cytosol*, a semifluid consisting mainly of water containing free-floating molecules.
Endoplasmic reticulum	This is a network of tubes fused to the membrane surrounding the nucleus and extending through the cytoplasm to the plasma membrane. It stores and separates substances and transports them through the cell's body. There are two kinds of endoplasmic reticulum. *Rough endoplasmic reticulum* is studded with *ribosomes* (described below) that make the proteins for the cell membrane or cell secretions. *Smooth endoplasmic reticulum* lacks ribosomes.
Ribosomes	These are protein factories that churn out proteins as dictated by the cell's genetic material. Every cell contains thousands of ribosomes. In fact, they make up one quarter of a typical cell's mass. There are two kinds of ribosomes. *Stationary ribosomes* are embedded in the rough endoplasmic reticulum; *mobile ribosomes* inject proteins directly into the cytoplasm.
Golgi apparatus	If ribosomes are protein factories, the Golgi apparatus is a protein packaging plant. Consisting of numerous layers of membranes that form a sac, this organelle contains special *enzymes* that modify, store, and transfer proteins from the endoplasmic reticulum to the cell membrane or other parts of the cell.
Lysosomes	These are the cell's waste-processing stations. They contain enzymes that help the cell digest proteins, fats, and carbohydrates; they also transport undigested material to the cell membrane for expulsion.
Vacuoles	These membrane-enclosed sacs serve many functions. Some store substances inside the cell, others help with digestion, and still others assist with waste removal. Plant cells typically have a large central vacuole that stores water and also plays important roles in the plant's reproduction, growth and development. Some one-celled organisms have vacuoles that can contract to expel water, giving them jet propulsion.

Table 2-1.   Cell Organelles *(Continued)*

Organelle Name	Description
Plastids	Found only in plants, these double-membraned structures contain pigments such as *carotinoids* (the stuff that makes carrots orange) and *chlorophyll* (the stuff that makes leaves green). Plastids that contain chlorophyll are called *chloroplasts*, and are home to *photosynthesis*, the process by which plants use sunlight to make sugar molecules for food.
Mitochondria	Mitochondria are the second-largest organelles in the cell, and have their own genetic structure (discussed in detail in Chapter 10). They have two membranes and inner folds, called *cristae*, which contain enzymes for the production of *adenosine triphosphate (ATP)*, the basic energy source for cells. Mitochondria also control the levels of water and other materials in the cell, and recycle and decompose proteins, fats, and carbohydrates, producing the waste product *urea* in the process.
Cytoskeleton	This network of fibers provides the cell with structural support, helping it keep its shape. It also aids the movement of organelles within the cell and the transport of materials into and out of the cell. The cytoskeleton consists of three sizes of filaments: *microfilaments,* thin solid rods that help some cells change shape; the fibrous *intermediate filaments;* and *microtubules,* the thickest, which are hollow. (The *flagella* and *cilia* that give some one-celled animals and certain other cells like sperm the ability to move are made up of microtubules.)
Centriole	Centrioles are cylinder-shaped organelles that are involved in cell division. Animal cells typically have a pair of centrioles lying at right angles to each other near the nucleus. Each centriole consists of nine tubes, each of which is in turn made up of three microtubules.
Nucleus	This is the compartment where the cell's genetic material is stored. It includes the *chromosomes,* composed of *DNA,* which contain the cell's genetic information, and the *nucleolus*—a spherical structure that disappears during cell division—which contains the RNA necessary for manufacturing proteins according to the instructions encoded in the DNA. The nucleus is surrounded by the *nuclear membrane*, composed of two layers with numerous openings in it to permit communication with the rest of the cell. Surrounding the nucleolus and the chromosomes is the *nucleoplasm*, which contains materials for building DNA and the *messenger RNA (mRNA)* that carries instructions between the nucleus and the cytoplasm.

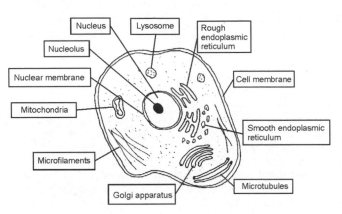

Fig. 2-1. This diagram shows the principal parts of an animal cell.

Counting Chromosomes

Better microscopes and better staining techniques also allowed scientists to see in detail what happened during cell division. In particular, early in the 19th century, they began to see, within the nuclei of both plant and animal cells, small structures that they called chromosomes (from the Greek *chromo,* "color," and *soma,* "body," because these tiny bodies became brightly colored when treated with certain dyes).

Only dead cells could be stained to show these chromosomes, but by careful ordering of cells that were stained at various points in the process of cell division, scientists were able to construct a kind of slide show of cells reproducing themselves.

By the second half of the 19th century researchers knew that all the cells of any individual member of any eukaryotic species (except for eggs and sperm) had the same number of chromosomes, and that number was characteristic of the species as a whole: Humans have 46, corn has 20, rhinoceroses have 84, and so on.

These chromosomes could be grouped into pairs based on their similar appearance under the microscope: humans have 23 pairs, corn has 10, rhinoceroses have 42, and so on.

The cell-division slide show assembled from micrographs of stained cells revealed that during the course of the process, each chromosome is duplicated, so that the total number of chromosomes doubles. Thus, each of the daughter cells produced during division ends up with the same number and kind of chromosomes as the parent cell.

CROSS-REFERENCE

We'll look at chromosomes in detail in Chapter 4, "Chromosomes—Organized DNA."

The Cell Cycle

The sequence of events from the start of one cell division to the next is called the cell cycle. Its length varies from cell to cell, but in a typical animal cell, it might be between 18 to 24 hours (see Fig. 2-2).

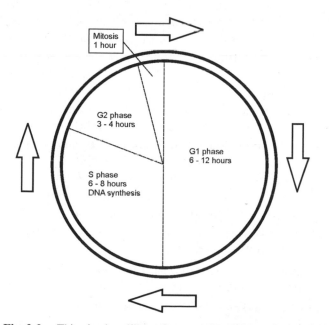

Fig. 2-2.   This pie chart illustrates approximately how long each phase of the cell cycle takes in a typical animal cell.

The cell cycle proceeds like this (see Fig. 2-3).

1. *Interphase.* This accounts for about 90 percent of the time in the cell cycle. It consists of three subphases.

   - G_1 *(gap1).* During this phase, the cell prepares to synthesize DNA. This typically lasts six to 12 hours.

   - *S (synthesis).* During this phase, the DNA molecules in each chromosome replicate themselves, becoming two identical DNA molecules called *chromatids*. Thin strands of *chromatin* (the easily stained complex of DNA and proteins that gives chromosomes their name) become visible under the microscope in prepared cells. This typically lasts six to eight hours.

   - G_2 *(gap 2).* During this phase, cell growth and expansion occur. G_2 typically lasts three to four hours.

2. *Mitosis.* During mitosis, the cell divides into two identical cells. This accounts for only one hour of the cell cycle, but it's a busy hour, consisting of four phases:

   - *Prophase.* In this phase, the chromosomes condense—rather like a rubber band—becoming shorter and more tightly wound as they are twisted. The DNA in the chromatin wraps itself around proteins, forming balls called *nucleosomes* like pearls on a necklace. This "pearl necklace" then spirals into a roughly cylindrical shape. Finally, the cylinder folds back and forth on itself, until what started as a long string of chromatin has been shortened to something only a few thousandths of its original length. At the end of this process, chromosomes have the short, thick appearance we're used to seeing in microphotographs, and consist of a pair of chromatids connected at their *centromeres* by a pair of *centrioles*, made of microtubules. Meanwhile, the cell itself is forming the *spindle*, a series of fibers made of microtubules, that extend from the poles of the cells from regions called the *centrosomes*. Some of the fibers run from pole to pole, while others connect to the chromosomes at a dense granular body called the *kintechore*, located in the same region as the centromere.

   - *Metaphase.* During this phase, the spindle fibers are shortened or lengthened as required to move the chromosomes into a plane, usually near the center of the cell, called the *metaphase plane*.

   - *Anaphase.* The chromatids separate at their centromeres and are pulled by the spindle fibers to the opposite poles of the cell.

- *Telophase.* The two sets of separated chromatids (two complete sets of chromosomes) assemble at the two poles of the cell, then begin to uncoil back into their string-like interphase condition. The spindle falls apart, the nuclear membrane reforms itself, and the cytoplasm divides (a process called *cytokinesis*). In animals, cytokinesis usually involves the formation of a furrow down the middle of the cell that eventually pinches it in two. In plants, cytokinesis usually involves the construction of a plate in the center of the cell that spreads out to the cell wall and then is strengthened into a whole new cell wall dividing the two daughter or progeny cells—each of which has exactly the same number and type of chromosomes as the previous cell.

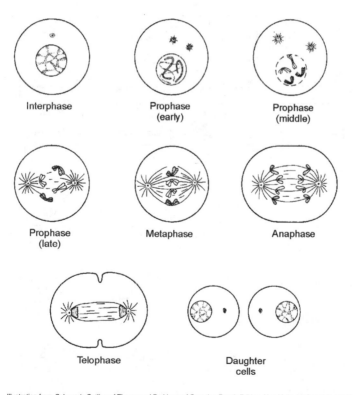

Illustration from *Schaum's Outline of Theory and Problems of Genetics*, Fourth Edition, New York: McGraw-Hill, 2002.

Fig. 2-3.   The stages of mitosis in animal cells. Maternal chromosomes are *light-colored* and paternal chromosomes are *dark.*

Connecting Mendel with Chromosomes

When Mendel's work was rediscovered at the turn of the 20th century, extensive studies of chromosome behavior had already been carried out. In 1902, American geneticist Walter Sutton (when he was just 25) was one of the first to propose that the "factors" Mendel had written about must be part of the chromosomes.

What convinced Sutton of that was the way in which the sex cells, the sperm and the egg, are created. As I've just described, ordinary cells, as they reproduce, create a complete copy of their chromosomes for their two daughter cells. Sex cells (*gametes*) are produced through a different process—not mitosis, but *meiosis*.

Meiosis is broken down into two divisions, called *meiosis 1* and *meiosis 2*. In meiosis 1, the chromosomes that duplicated themselves during interphase

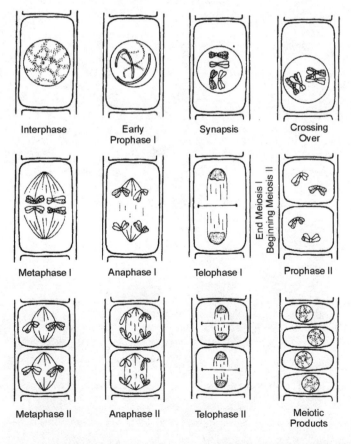

Illustration from *Schaum's Outline of Theory and Problems of Genetics*, Fourth Edition, New York: McGraw-Hill, 2002.

Fig. 2-4. The stages of meiosis in plant cells. Maternal chromosomes are light-colored and paternal chromosomes are light.

thicken and condense just as they do during mitosis. But instead of lining up on a plane down the middle of the cell, they lie side by side to form pairs called *bivalents,* each of which contains all four of the cell's copies of one chromosome. These bivalents divide into chromatid pairs, generating two sets of chromosomes.

Then, in meiosis 2, both of the resulting sets of chromosomes divide again, with one member of each resulting set migrating to a new daughter cell. The end result is that the starting number of *4n* chromosomes is distributed among four different cells, each containing *n* chromosomes. These *haploid cells* become gametes (eggs or sperm). (See Fig. 2-4.)

When a sperm fertilizes an egg, the resulting *zygote* thus receives half of its chromosomes from one parent and half from the other parent.

This matched perfectly with Mendel's theory of factors, and convinced scientists that the ultimate basis of heredity must lie within the chromosomes. It was clear, however, that each chromosome had to carry more than one factor. Even in species with more chromosomes than humans, there obviously weren't enough to describe all the traits that made each individual unique.

The ultimate medium of genetic information had to be something much smaller and more complex. And so it turned out to be, as we'll see in the next chapter.

Quiz

1. Who wrote *Micrographia,* detailing his observations through the microscope?
   (a) Captain Cook
   (b) Captain Hook
   (c) Robert Hooke
   (d) Robert Schuck

2. The distinct spot Robert Brown observed near the center of orchid cells is called the
   (a) core.
   (b) heart.
   (c) heredicolus.
   (d) nucleus.

3. Schleiden and Schwann are considered the fathers of
   (a) hot dogs and bicycles.
   (b) cell theory.
   (c) genetic manipulation.
   (d) eugenics.

4. Cells containing a well-defined nucleus and other membrane-bound structures are called
   (a) eucalyptus cells.
   (b) euphemism cells.
   (c) Eustace cells.
   (d) eukaryotic cells.

5. Cells that have only a single outer membrane and lack a defined nucleus and other internal structures are called
   (a) antinuclear cells.
   (b) proficient cells.
   (c) prokaryotic cells.
   (d) nonlinear cells.

6. The smaller structures within cells are called
   (a) organelles.
   (b) pianonelles.
   (c) nervousnelles.
   (d) protonelles.

7. Mitosis refers to
   (a) staining cells so their structures can be viewed under the microscope.
   (b) growing cells in a Petri dish.
   (c) the duplication and division of a cell's chromosomes.
   (d) bad breath.

8. The spindle is
   (a) a laboratory instrument used to study cell growth.
   (b) a fibrous structure that positions and divides chromosomes during cell division.
   (b) the point at which a pair of chromosomes are joined.
   (d) a sharp knife used to gather cell samples from animals.

9. The cell-division process that produces gametes (sperm and eggs) is called
   (a) cirrhosis.
   (b) pertussis.
   (c) myalis.
   (d) meiosis.

10. A cell that contains only half the full complement of chromosomes is said to be
    (a) haploid.
    (b) paranoid.
    (c) alloyed.
    (d) celluloid.

CHAPTER 3

DNA—The Chemical Basis of Heredity

In 1869, just three years after Mendel's paper was published, a Swiss chemist named Johann Friedrich Miescher made a discovery. When pure cell nuclei, obtained from pus cells collected from bandages discarded by hospital surgical wards, were treated with weakly alkaline solutions which were then neutralized by acids, an odd substance precipitated out.

It was thought that animal cells were largely made of protein, but the new substance, Miescher stated (and confirmed by chemical tests) "cannot belong among any of the protein substances known hitherto."

Since it was contained in the cell nuclei, he dubbed the substance "nuclein." Nuclein was soon discovered in other types of cells. Chemical analysis showed that in addition to hydrogen, carbon, nitrogen, and oxygen—elements common to organic molecules—nuclein also contained phosphorous.

Miescher's paper announcing the discovery of nuclein appeared in 1871. He theorized that nuclein was a storehouse of phosphorus waiting for use by the cell for some unknown purpose.

In truth, Miescher had become the first person to see deoxyribonucleic acid, or DNA, the molecule whose unique properties are central to the process of heredity.

The Race to Decipher DNA

The importance of Miescher's discovery remained unknown for more than half a century. As research continued into the biochemical basis of genetics in the early part of the 20th century, scientists began to accept that a gene had to be some kind of complex chemical. They knew that genes were to be found in chromosomes (more on those in the next chapter). They also knew that the only two complex chemicals that chromosomes contained were proteins and the nucleic acids (as nuclein had been renamed since Miescher's time). These were DNA and a related chemical, ribonucleic acid (RNA).

Most geneticists thought nucleic acids weren't complicated enough to carry the genetic code, and so the conventional wisdom was that genes would turn out to be some kind of proteins. All that was turned on its head by experiments conducted in the late 1930s and early 1940s by Oswald Avery, an immunologist at Rockefeller University in New York.

A British Ministry of Health researcher named Fred Griffith had shown earlier that a nonvirulent form of *Streptococcus pneumoniae* bacteria, called *rough,* could be transformed into the virulent form, called *smooth.* Griffith injected live rough cells and dead smooth cells into mice. Within two days, many of the mice had died. From their blood, Griffith recovered live smooth cells. *Something* had transformed the rough, nonvirulent form of the bacteria into the smooth, virulent form.

Over a decade, in a series of painstaking experiments, Avery worked to discover what this "transforming principle" might be. His conclusions defied expectations: the transforming principle of *Streptococcus pneumoniae*, and by extension all forms of life, was DNA.

Avery's results weren't immediately accepted by everyone (scientific discoveries seldom are), but over the next few years, further experimentation confirmed the centrality of DNA to genetics. In particular, research by Alfred Hershey and Martha Chase showed that the genetic information viruses inject into cells to infect them is carried by DNA, not by the viral coat. (What was the high-tech equipment they used to separate viruses' DNA loads from their coats? A kitchen blender.)

Avery's discovery raised another question: If DNA carried genetic information, how did it do so? Scientists reasoned the answer must lie in the molecule's structure—and so the race was on to decipher that structure.

Three groups of researchers began searching for the structure of DNA even before Hershey and Chase's results were published. One group, at Caltech in the U.S., was headed by chemist Linus Pauling. Two groups were in England. One, at King's College in London, was led by Maurice Wilkins. One of his colleagues was chemist Rosalind Franklin, whose particular area of expertise was X-ray crystallography (*see sidebar*). The second group, at Cambridge, included American James Dewey Watson and Englishman Francis Harry Compton Crick.

Through thinking, talking, and model building, Crick and Watson attempted to decipher DNA's structure, with little success. But on January 30, 1953, Watson visited Maurice Wilkins at King's College. The two had met several years before and had become friends. Wilkins showed Watson "Photograph 51," an X-ray photo of a DNA crystal made by Rosalind Franklin (who had not given her permission that it be shared). On viewing that photograph, Watson had a flash of inspiration. He told Crick his thoughts, and over the next month they refined their vision of the structure of DNA. Their short paper on the subject appeared in the British science journal *Nature* on April 25, 1953. Five weeks later they published a second paper which explained how DNA's structure was the key to its method of self-replication—the basis of reproduction of all forms of life on earth.

Rosalind Franklin, 1920–1958

Rosalind Franklin decided to become a scientist when she was 15, and went on to attend Cambridge. (Her father initially refused to pay for her education because he disapproved of university education for women; an aunt stepped in instead.)

After graduating in 1941, Franklin began studying coal and charcoal and how to use them efficiently; her still-quoted papers helped launch the field of high-strength carbon fibers. At age 26, with a brand new PhD, she began to work in x-ray diffraction—the use of x-rays to study the molecular structure of crystals. She expanded the use of the technique from the study of simple crystals to the study of complex, unorganized matter such as large biological molecules.

Her work led to an invitation in 1950 to join a team of scientists at King's College in London who were studying living cells. She was assigned to work on DNA. Unfortunately, she didn't get along with Maurice Wilkins, the laboratory's second-in-command and the one leading the DNA work.

By careful adjustment of her equipment to produce an extremely fine beam of x-rays, and better preparation of her DNA specimens, Franklin was able to produce astonishingly clear x-ray diffraction images of DNA's molecular structure. It was one such photo, shared without her knowledge with

Watson by Wilkins, that helped Watson and Crick successfully decipher the structure of DNA before the King's College team did.

Shortly after that, Franklin left King's College (on the condition she not work on DNA). She returned to her study of coal and turned her attention to viruses. Her research group at Birkbeck College in London laid the foundation for structural virology.

Franklin died of ovarian cancer in 1958 at age 37. Many people believe she deserved a share of the 1962 Nobel Prize won by Watson, Crick, and Wilkins. However, the Nobel Prize criteria do not allow for posthumous recognition of scientists.

The Double Helix

Even before Watson and Crick figured out the structure, scientists realized that DNA molecules have a backbone whose "vertebrae" are units consisting of one molecule of a sugar (deoxyribose) and one of phosphate (a compound that contains phosphorus).

Attached to each vertebra is a small molecule called a *nitrogenous base*, consisting of oxygen and nitrogen atoms arranged in rings. There are four types of these: *adenine* (abbreviated A), *guanine* (G), *cytosine* (C) and *thymine* (T). Adenine and guanine, whose ring-shaped molecules have six members, are called *pyrimidines*; cytosine and thymine, with fused five- and six-member rings, are called *purines* (see Fig. 3-1).

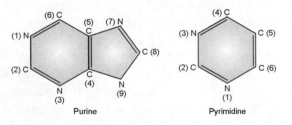

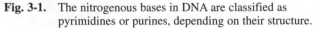

Fig. 3-1. The nitrogenous bases in DNA are classified as pyrimidines or purines, depending on their structure.

In the late 1940s Austrian-born chemist Erwin Chargaff showed that in any DNA molecule, the amount of cytosine equals the amount of guanine and the amount of adenine equals the amount of thymine.

X-ray crystallography hinted that DNA might have a helical or spiraling, structure, as do many proteins. Watson and Crick's breakthrough was realizing that DNA consists of two backbones whose bases face inward toward each other and interlock like the rungs on a ladder (see Fig. 3-2). Connected by weak chemical bonds called *hydrogen bonds,* adenine always pairs with thymine and cytosine always pairs with guanine—which explains Chargaff's discovery.

Because the same bases always pair up, one helix of the DNA molecule is always a mirror image of the other. Watson and Crick realized that if you split DNA down the middle in an environment in which there were free-floating backbone units and bases (such as the nucleus of a cell), both sides would combine with those free-floating units to create a perfect copy of the original molecule.

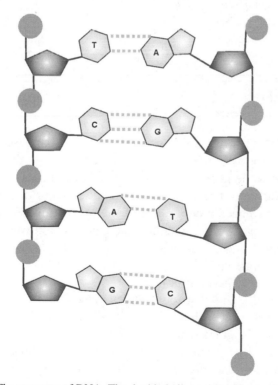

Fig. 3-2.   The structure of DNA. The double helix maintains a constant width because of the pairing of thin pyrimidines with thick purines.

So, now you would have two identical molecules, perfect twins of the original, parent molecule. The fact a thick purine always pairs with a thinner pyrimidine ensures that the two backbones remain at a constant distance from each other.

The weak hydrogen bonds between the bases don't hold the molecule together; that duty is performed by the much stronger bonds between deoxyribose rings and the phosphate groups in the backbones. Each deoxyribose is connected to two different phosphates. Each of these links is to a different carbon atom on the deoxyribose ring, one of which is labeled 3′ (read as "three-prime"), and the other of which is labeled 5′. This gives each of the backbone strands of DNA directionality. Reading up the left-hand strand or down the right-hand strand is referred to as reading in the 3′-to-5′ direction, while reading down the left-hand strand or up the right-hand is reading in the 5′-to-3′ direction. (This directionality plays an important role in DNA replication.)

Each base pair is rotated approximately 36° around the axis of the helix relative to the next base pair, so there are typically 10 base pairs in each complete turn of the helix.

Watson, Crick, and Wilkins shared the Nobel Prize in Physiology or Medicine in 1962 for the discovery of DNA's structure.

Replication of DNA

To replicate, DNA must first split apart. The fact that the bases are joined by relatively weak hydrogen bonds is important. This allows the molecule to split down the middle while keeping the outer backbones intact.

Even so, replication of DNA obviously entails major disruption of the DNA molecule. If the two backbone strands of the original molecule separated completely, they would probably fragment and the genetic information would become scrambled. Fortunately, the two backbones of the original molecule never exist as completely separate strands. Instead, only a small part of the DNA loses its duplex structure at any given moment.

The point at which this disruption takes place is called the *replication fork*. The replication fork moves along the parental DNA which unwinds, then winds again, into the two daughter strands (see Fig. 3-3).

This complex process requires the action of at least three *enzymes*. Enzymes are proteins that allow specific chemical reactions to occur more easily, without themselves being changed by the reaction. (The suffix -ase typically denotes an enzyme.)

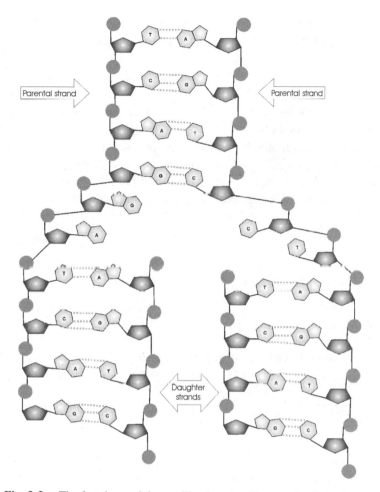

Fig. 3-3.   The fact that each base will only pair with one other base allows DNA to faithfully replicate itself.

Enzymes called *helicases* first open the double helix, producing single-strand templates. Binding proteins then stabilize these regions to form the replication fork.

Enzymes called *DNA polymerases* do the work of building two copies of the original DNA molecule by attaching *nucleotides* to the exposed bases. A nucleotide is one piece of DNA backbone. It is a sugar/phosphate pair with a base attached. Interestingly, polymerases can not only add free nucleotides to existing chains, they can remove them. During replication, incorrectly paired bases are very likely to be removed by polymerases and replaced with the proper ones. This helps prevent the introduction of errors (mutations) into the daughter DNA strands. (More on DNA's self-repair mechanisms a little later in this chapter.)

All known DNA polymerases can attach nucleotides only to the 3′ ends of the exposed DNA backbones, so only one strand—the one with a free 3′ end—can be replicated continuously. This is called the *leading strand*. The other strand, the *lagging strand*, is replicated in short pieces called *Okazaki fragments*, each a few hundred nucleotides long, in the opposite direction to the movement of the replication fork. Gaps created by missing nucleotides and nicks between the fragments are rapidly filled in by yet another kind of enzyme called a *DNA ligase*.

In a very long double helix, replication may actually start at more than one site at once. These sites are called "origin of replication" or *ori* sites.

DNA Mutation and Repair

DNA is very good at replicating itself accurately, but errors can creep in. Any change in the sequence of DNA from one generation to the next is called a *mutation*. Mutations can occur anywhere in the genetic sequence, but not all mutations result in physical changes to the organism. Such changes are only seen if the mutation occurs in the sequence of a gene.

Damage to DNA is not necessarily permanent. Cells have a number of DNA repair mechanisms.

1. *Photoreactivation.* Some cells, such as the bacterium *E. coli*, contain a light-dependent enzyme that can remove or correct damaged DNA.
2. *Excision.* This is a four-step process:
   a. One strand is broken by an enzyme called *UV endonuclease*.
   b. A DNA polymerase removes nucleotides near the cut, including the damaged area.
   c. A DNA polymerase replaces the nucleotides with the correct ones, working from the uncut complementary strand.
   d. An enzyme called *polynucleotide ligase* seals the break.
3. *Mismatch repair.* Bases that aren't paired with their proper counterparts can also be repaired using excision.
4. *SOS repair.* Breaks in a DNA strand can sometimes simply be sealed without any regard for the original sequence. This is seen in cases of extreme DNA damage. While this may allow the cell containing the damaged DNA to survive, it greatly increases the possibility that the cell will contain one or more mutations.

CROSS-REFERENCE

Mutations and their effects on organisms are discussed in detail in Chapter 7.

Within cell nuclei, DNA molecules are organized into much larger bodies called *chromosomes*. Because chromosomes are easy to see in ordinary microscopes, they were actually studied in some detail long before the structure and function of DNA was understood.

In the next chapter we'll take a look at the structure of chromosomes, and their role in genetics.

Quiz

1. In what substance did Johann Friedrich Miescher first isolate DNA?
    (a) pond water
    (b) urine
    (c) pus
    (d) egg yolk

2. What are DNA and RNA collectively called?
    (a) central vacuoles
    (b) nucleic acids
    (c) fissionable molecules
    (d) hereditons

3. Whose work confirmed DNA as the "transforming principle" of genetics?
    (a) Avery Brooks
    (b) Tex Avery
    (c) Oswald Avery
    (d) Avery Johnson

4. What photographic technique did Rosalind Franklin apply to DNA?
    (a) polarizing filters
    (b) microphotography
    (c) x-ray diffraction
    (d) time-lapse imaging

5. The backbones of the DNA double helix are made of
   (a) sugars and phosphates.
   (b) protein and calcium.
   (c) acids and alkalis.
   (d) salts and metals.

6. Which of these is not one of the nitrogenous bases of DNA?
   (a) adenine
   (b) piscine
   (c) guanine
   (d) thymine

7. One piece of DNA backbone with a base attached is called a
   (a) nucleotide.
   (b) nucleosome.
   (c) chromotide.
   (d) chromosome.

8. The point at which a DNA molecule splits apart is called a
   (a) reproductive split.
   (b) copying center.
   (c) replication fork.
   (d) fission focus.

9. Proteins that act as catalysts, allowing biochemical reactions to take place without themselves being changed, are called
   (a) ribosomes.
   (b) chromosomes.
   (c) cytosines.
   (d) enzymes.

10. Any change in the sequence of DNA from one generation to the next is called a
    (a) crustacean.
    (b) mutation.
    (c) extrusion.
    (d) confusion.

4

Chromosomes—Organized DNA

Long before anyone had figured out the connection of DNA to heredity—much less deciphered its molecular structure—heredity had been linked to a particular part of the cell.

Walther Flemming, a German scientist, discovered in 1875 that the nucleus of cells contained a substance that could be easily stained with dye. He called it *chromatin,* from the Greek word for color. Flemming noted that just before a cell divided, the chromatin formed stringy bodies, which he called *chromosomes.*

As Mendel's theories became more accepted, scientists began looking inside cells for his mysterious "factors." In 1902, American geneticist Walter Sutton was one of the first to suggest that perhaps the factors were somehow connected to the chromosomes. He'd noted that as cells divided, the chromosomes grouped themselves into pairs, which first duplicated themselves, then pulled apart, so that each daughter cell ended up with a set of chromosomes just like those in the parent cell.

He also noted that when sperm and egg cells formed, the pairs of chromosomes split, but did not duplicate themselves. That meant that each sex cell ended up with only half the normal number of chromosomes. This, in turn, meant that when a sperm fertilized an egg, the resulting organism got half its chromosomes from one parent and half from the other—which was just the way Mendel's factors worked.

The Fly Room

Another American geneticist, Thomas Hunt Morgan, set out to test Sutton's theory. Morgan actually doubted Sutton was right. After all, he pointed out, every cell in the body had the same chromosomes, yet the cells themselves were very different—muscle cells were little like blood cells which were little like brain cells.

In 1904, Morgan began working at Columbia University in New York City. He and his students used a variety of animals in their studies of genetics, but they found the fruit fly—*Drosophila melanogaster*—the most useful. For one thing, they could be (and were) easily lured into the laboratory simply by placing bananas on the windowsill, and can be kept quite happily in nothing more exotic than a milk bottle. For another, they were small enough—about a quarter of an inch in length—so that thousands could be kept in a small laboratory. And best of all, from a geneticist's point of view, they reproduced quickly, producing a new generation every 12 days, in vast numbers (the average female lays about 1000 eggs). Morgan's laboratory soon became known as "The Fly Room."

One day in 1910, Morgan spotted a male fly whose eyes were white rather than the normal red. He dubbed the trait, rather unimaginatively, white, and proceeded to breed the white-eyed male with a red-eyed female. All the progeny had red eyes. But when he crossed those flies with each other, the white-eyed trait reappeared—but only in the males. Somehow, it was "sex-limited," occurring only in one gender.

Morgan knew that fruit flies have four pairs of chromosomes, including a pair of sex chromosomes. In 1905, Edmund Wilson and Nettie Stevens had discovered that in many animals—including both fruit flies and humans—males had one chromosome in one pair that was smaller than the other and a different shape. They dubbed it the Y chromosome. Females inherited two normal-looking chromosomes, which were called X chromosomes, while males got an X chromosome from their mother and a Y chromosome from their father.

Morgan deduced that white, and other such sex-limited traits—such as human red/green color blindness, which only afflicts men—must reside on the X chromosome (see Fig. 4-1).

That was the first mutation to be linked to a specific chromosome, and the first solid evidence that Mendel's factors were indeed to be found in the chromosomes. Over the next few years, Morgan and his coworkers identified 40 different mutations in fruit flies—not just different eye colors, but differently shaped body parts. Some of the mutations seemed to be linked together, and those linkages were clustered into four groups. Since there were four chromosome pairs in the fruit fly, this was further evidence that Mendelian factors were indeed to be found in the chromosomes.

Many other discoveries about chromosomes flowed out of Morgan's lab at Columbia (I'll point them out as they come up throughout this chapter). Morgan himself won the Nobel Prize for Physiology and Medicine in 1933 for his discoveries concerning the heredity function of the chromosome.

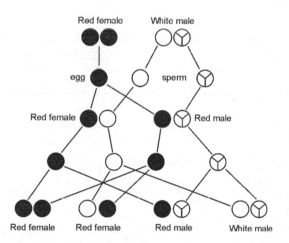

Fig. 4-1.   This chart shows the result of crossing a red-eyed female and a white-eyed male fruit fly. Black circles are chromosomes which carry the red-eyed trait. White circles are chromosomes carrying the white-eyed trait. White squares with Ys in them are Y chromosomes carrying the white-eyed trait. The red trait is dominant, so females will always have red eyes, because they always receive at least one red-eyed X chromosome from their mother. However, they can carry the white-eyed trait on their second X chromosome. Males who inherit that X chromosome along with the white-eyed Y chromosome will have white eyes.

The Structure of the Chromosome

Now, of course, thanks to a better understanding of the molecular basis of heredity, we also have a better understanding of the chromosome than Morgan and his coworkers could hope for.

Each chromosome consists of DNA and protein bundled together. The DNA is tightly wound and folded on a protein core. How tightly? Well, the DNA in a typical human cell has to fit inside a nucleus only 0.005 millimeter (.0002 of an inch) in diameter. If you could fully extend it, it would stretch to almost two meters, or around six feet. You could put about 200 human cells in the period at the end of this sentence, so a dot the size of a period could contain 400 meters of DNA. There are 100 trillion cells (approximately) in the human body, so the DNA in your body alone, if stretched out in a single fine thread, would run to the sun, 93 million miles away, and back—20 times or so. And despite being so tightly packaged to begin with, the genetic material in a cell becomes even more tightly packaged—five to 10 times more—as a cell begins to divide (as described in Chapter 2). This is when chromosomes become readily visible with staining.

As also noted in Chapter 2, different species have different numbers of chromosome pairs. (The number of chromosomes in any given cell is called its *karyotype*.) In humans, there are 23 pairs. The first 22 pairs, called the *autosomes,* are labeled according to length, longest to shortest, 1 through 22. The 23rd pair are the sex chromosomes—the X and Y chromosomes mentioned earlier.

The *centromere,* vital to the proper alignment of the chromosome pairs before cell division, divides the chromosome into two arms of varying lengths. Its location varies from chromosome to chromosome. If it's near the center, the chromosome will have two arms of almost equal length and is called *metacentric.* If it's slightly off-center, one arm will be longer than the other, and the chromosome is *submetacentric.* If it's greatly off-center, the size difference between the arms will be even greater, and the chromosome is called *acrocentric.* In those cases, the shorter arm is called the *p* arm and the longer arm is called the *q* arm. Finally, if the centromere is at or very near the end of the chromosome, the chromosome is called *telocentric* (see Fig. 4-2).

Also at the ends of the chromosomes are structures called *telomeres* which, rather like plastic or metal aglets that finish off the tips of shoelaces, seal the chromosome's ends. Discovered by Nobel laureate Hermann Muller in the 1930s, telomeres, which consist of thousands of repeating bits of DNA strung together, help the cell recognize the difference between the end of a chromosome and a broken chromosome that needs repair. The telomeres get shorter and shorter as a cell divides, and seem to play a role in programming the death of the

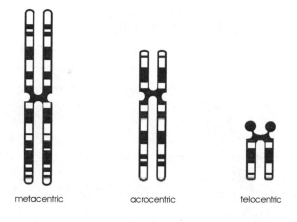

metacentric   acrocentric   telocentric

Fig. 4-2.   Chromosomes can have different appearances,
depending on where the centromere is located.

cell. One of the distinguishing characteristics of cancer cells is that they activate
an enzyme called telomerase, which keeps the telomeres from shrinking and
makes the cells effectively immortal.

Linkage, Crossing Over, and Chromosome Mapping

As mentioned earlier, Morgan and his students found that certain traits were
linked. That is, the genes for those traits were located on the same chromosome.

To determine linkage, Morgan's team would first breed any new fly mutant
they found with a normal fly. The offspring of that mating were then mated with
each other, over and over again, until the researchers had created a dependable
stock of flies showing the mutation.

Males with the new mutation were then mated with females carrying other
mutations already known to be on a specific chromosome. Because those mutations had already been located on a specific chromosome, they served as markers.
White eyes, for instance, were produced by a mutation in the X chromosome. If,
in succeeding generations, the new mutation consistently appeared in flies with
the white-eye mutation, the researchers would know that the X chromosome also
carried the gene for the new mutation.

Mendel Was Lucky

As you'll recall, one of Mendel's principles of inheritance was the principle of independent assortment. That is, each of the traits he studied in his pea plants was independent of the others, so that whether a pea plant had one of them (say, wrinkled seeds) had no effect on whether it had another of them (say, tallness).

Morgan found that Mendel had been lucky. The seven traits he chose to study were all carried on different chromosomes of the pea plant—which only has seven pairs. Had some of the traits he studied been linked, his results would not have been so clear-cut and he might never have come up with his seminal theories of inheritance.

The concept of linkage leads to another important discovery. Morgan's team was puzzled by a mutation called *miniature wing*. It was linked to white eye (genes for both traits belonged to the X chromosome) so the two were usually inherited together, but not always.

Morgan came up with an explanation. F. A. Janssen, a Danish biologist, had discovered that just before dividing while forming sex cells, pairs of chromosomes twisted around each other. Morgan suspected that when the chromosomes then separated, they broke into pieces that rejoined to form whole chromosomes again. In that breaking and rejoining process, part of one chromosome could stick onto the other. All of the genes on that broken bit would therefore move from one piece to the other. Morgan called this *crossing over*. It's now also known as *recombination*.

This could explain why miniature wing was not reliably inherited with white eye. Although they were linked (that is both belonging to the same chromosome), if they were rather far apart on the chromosome, any crossing over during the formation of the sex cells could separate them, so that only one of them made it into one of the new cells. Linked genes that were seldom separated by crossing over would be those that were closer together on the chromosome (see Fig. 4-3).

In 1911, Morgan explained this idea of crossing over to his student Alfred Sturtevant. Sturtevent, who was still an undergraduate, realized that if Morgan was correct, it should be possible to roughly map the location on the X chromo-

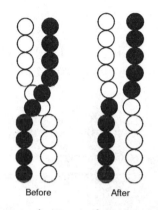

Before   After

Fig. 4-3.   When eggs and sperm are formed, pairs of chromosomes
separate, so that each sex cell has only half as many
chromosomes as a tissue cell. Just before this separation
happens, the chromosomes in each pair twist together.
Sometimes they break, and part of one chromosome gets
stuck onto the other. The result is a chromosome that
contains genes from both members of the original pair.
This is called *crossing over* or *recombination.*

some of the genes they had already identified as belonging to it. This would be
based on how often the various traits showed up in the progeny resulting from
the mating of flies with pairs of sex-linked traits. As he later wrote, "I went home
and spent most of the night (to the neglect of my undergraduate homework) in
producing the first chromosome map." See Fig. 4-4.

Sturtevant and others on Morgan's team built maps of three other chromosomes, and first set forth the idea that some traits arose through the combined
actions of several different genes, and that some genes could modify the actions
of others.

Their early chromosomal maps were the ancestors of all the genetic maps that
have been constructed since, right up to the Human Genome Project.

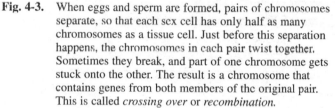

Fig. 4-4.   The first genetic map looked something like this, showing the
locations of five traits on the X chromosome of a fruit fly.

CROSS-REFERENCE

Genome mapping is discussed in more detail in Chapter 6, and there's more information about the Human Genome Project in Chapter 12.

Crossing Over Confirmed

Morgan and his team were convinced crossing over occurred, but they weren't able to observe it directly—chromosomes were simply too small to make out the kind of detail direct observation required.

That changed in the 1930s, when scientists discovered that cells in the salivary glands of fruit fly larva contained giant chromosomes, 2000 times bigger than normal ones. Not only that, the giant chromosomes had stripes—light and dark bands—that were easy to recognize and count. For the first time, researchers could directly observe crossing over and other chromosomal activity Morgan and his team had only been able to infer—and observation confirmed they had inferred correctly.

Identifying the genetic material and understanding how it is organized inside living cells solved only one part of the genetic puzzle.

It's one thing to understand that DNA carries genetic information. But how is that information used to build a living creature? How does something as tiny as a DNA molecule, no matter how complex it may be, exert control over physical characteristics on a vastly larger scale?

That's the topic of the next chapter.

Quiz

1. The substance chromosomes are made of was called chromatin
   (a) because it contains chrome.
   (b) because it contains tin.
   (c) because that was the name of its discoverer.
   (d) because it can be easily colored for observation.

2. Compared to body cells, how many chromosomes do sex cells have?
   (a) The same number.
   (b) Half as many.
   (c) Twice as many.
   (d) One quarter as many.

3. What insects did Thomas Morgan and his students use to research genetics at Columbia University?
   (a) House flies.
   (b) Horse flies.
   (c) Fruit flies.
   (d) Butterflies.

4. What are the sex chromosomes labeled?
   (a) Male and female.
   (b) Mars and Venus.
   (c) Gamma and epsilon.
   (d) X and Y.

5. What is a "sex-limited" trait?
   (a) One that only affects reproductive organs.
   (b) One that occurs only in one gender.
   (c) One that renders the resulting organism sterile.
   (d) One that occurs only in organisms that don't have sex.

6. How many pairs of chromosomes do humans have?
   (a) 23
   (b) 32
   (c) 12
   (d) 7

7. What divides chromosomes into two arms of varying lengths?
   (a) The telomere.
   (b) The ventricle.
   (c) The dorsal pin.
   (d) The centromere.

8. If two genes are linked, they
   (a) appear on the same chromosome.
   (b) form a connection between two different chromosomes.
   (c) directly affect one another.
   (d) prevent a chromosome pair from dividing properly.

9. What Morgan called "crossing over" is also called
   (a) channeling.
   (b) bridging.
   (c) recalibration.
   (d) recombination.

10. The first gene maps revealed
    (a) where specific genes resided on the chromosome.
    (b) where chromosomes resided in the cell nucleus.
    (c) where sex cells came from.
    (d) how to get to Columbia University's "Fly Room."

Traits—How Genes Are Expressed

Francis Crick was one of those who wanted to understand how the information encoded in DNA was expressed in the physical structure of the organism. One hint came from sickle-cell anemia, a genetic disorder that results in abnormal red blood cells that have a greatly shortened lifespan and that afflicts about 0.2 percent of all babies of African descent born in the United States. By the early 1940s, scientists had discovered that sickle-cell anemia's inheritance pattern follows Mendel's rules for a single gene. In the 1950s scientists determined that the mutation that causes sickle-cell anemia changes a single amino acid in one of the polypeptide chains making up hemoglobin, the protein in red blood cells that carries oxygen to the body's tissues.

Drawing on this and other research, in 1957, Crick suggested that the order of the bases in a DNA molecule could be read as a code for the sequence of amino acids in protein molecules. If each set of three bases represented one amino acid, there would be 64 possible combinations. Since every living thing on earth is made up of the same 20 amino acids, that system of coding was plenty complex enough.

But which combinations of bases produced which amino acids? When Crick proposed the three-letter structure of what became known as the *genetic code,* that seemed an indecipherable mystery. But just a few years later, the entire code had been cracked.

Experiments with *E. coli* bacteria provided the solution. During the late 1950s, scientists found that they could synthesize polypeptides by introducing individual amino acids to a solution containing broken *E. coli* cells. In 1961, a young scientist from the National Institute of Health, Michael Nirenberg, demonstrated that adding RNA molecules containing only one kind of base, uracil, to a similar solution resulted in a polypeptide containing only one amino acid, phenylanine. That seemed to indicate that the three-base combination UUU was the *codon* (as the coding units became known) for phenylanine—and it also suggested a way to go about discovering the codons for other amino acids.

Over the next couple of years, other RNA molecules—first with randomly arranged base sequences and then with defined base sequences—were used in similar experiments to pin down the relationship between specific codons and specific amino acids. By 1964, the genetic code had been deciphered. See Fig. 5-1.

As you've probably noted, the code was cracked by adding RNA, not DNA, to the *E. coli* solution. That's because RNA is crucial to the way cells use the information contained in DNA to produce amino acids and, ultimately, proteins.

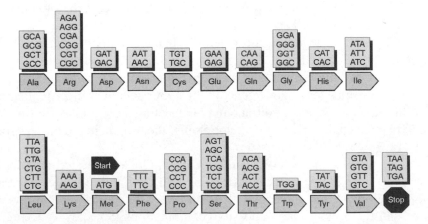

Fig. 5-1. The genetic code proved to be remarkably straightforward. Most amino acids can be coded for by more than one three-base combination, or codon; three codons serve as *stop codes,* signaling the place where construction of a polypeptide chain should end.

Transcription

The transformation of intangible genetic information into the tangible proteins that make up you, me, the cat, and every other living thing begins with transcription—the copying of the sequence of nucleotides in DNA onto a chain of RNA called *messenger RNA*, or *mRNA* for short.

Recall that RNA is very much like DNA, except that in its backbones it has the sugar ribose instead of the sugar deoxyribose, and the base uracil takes the place of the base thymine.

Two Ways to Write Codons

Because RNA contains the base uracil in place of the base thymine, there are two ways to write many of the codons in the genetic code.

If the subject is the codons in DNA, you'll see three-letter combinations involving the letter T, for thymine (e.g., TGG, TAT, TAC, etc).

If the subject is the codons as transcribed into mRNA, the letter U, for uracil, replaces the letter T. So the same codons become UGG, UAU, UAC, and so forth.

The process of transcription is carried out by enzymes called *RNA polymerases*. These polymerases begin by unwinding a portion of DNA near the start of a gene, which is defined by special nucleotide sequences called *promoters*. Only one of the strands of DNA contains actual codons. It's called the *coding strand* or *sense strand*. The complementary strand is called the *anti-coding* or *anti-sense strand*. Somewhat counter-intuitively, though, it's actually the antisense strand that mRNA latches on to in order to reconstruct the codons that occur in the sense strand. See Fig. 5-2.

Transcription moves in the 5′ to 3′ direction. The beginning of the code for a specific polypeptide chain is always indicated by the *start codon* AUG. Since that's also the codon for the amino acid methionine, all proteins begin with methionine. The end of the polypeptide sequence is indicated by the *stop codons* UAA, UAG, or UGA. Since none of the stop codons code for an amino acid, the last amino acid of the polypeptide chain is specified by the codon just before

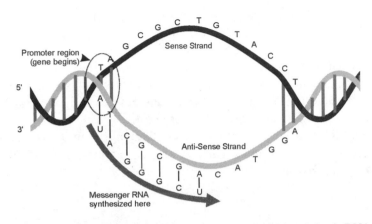

Fig. 5-2. In transcription, the codons in one strand of an organism's DNA are copied into a strand of mRNA.

the stop codon. (Transcription actually begins before the start codon and continues past the stop codon, so mRNA usually contains several extra nucleotides at both ends that do not code for amino acids.)

Translation

DNA normally cannot leave the cell nucleus; RNA can. The mRNA, carrying its amino acid codes, moves out from the nucleus into the cytoplasm, where it encounters small structures called *ribosomes*. Ribosomes consist of more than 50 different proteins and three or four different kinds of RNA called *ribosomal RNA (rRNA)*, and free-floating bits of yet another type of RNA, *transfer RNA (tRNA)*. tRNA functions rather like a fleet of tugboats, hauling amino acid molecules from place to place. Each different tRNA has a three-piece nucleotide sequence called an *anti-codon* that complements one of the codons on mRNA. Each tRNA also carries with it a molecule of the appropriate amino acid for its anti-codon. (A tRNA carrying an amino acid molecule is said to be *activated* or *charged*.) Attachment of each of the amino acids to the appropriate tRNA is carried out by an enzyme called *amino-acyl synthetase*.

The first difficulty in translating a strand of mRNA into the correct sequence of amino acids is deciding where to start reading. The same sequence can be read three different ways, depending on whether the first base of the first codon is taken to be the first, second, or third base of the string. In other words, a string that begins CUAUGGCAA... could be read as CUA UGG CAA... or UAU GGC AA... or AUG GCA A....

That's where the start codon, AUG, comes in. A ribosome, and a bit of tRNA called an *initiator* carrying a molecule of the amino acid methionine, binds to the mRNA's start codon. Then another piece of tRNA comes along bearing a molecule of whatever amino acid the second codon of the mRNA represents.

The ribosome aligns everything so the appropriate reactions can take place, and bonds the methionine molecule to the second amino acid molecule using an enzyme called *peptidyl transferase*. At the same time, the bond between the amino acid molecules and their tRNA "tugboats" is severed and the tRNA is released to go pick up another amino acid molecule. The ribosome moves to the next codon and the process repeats until all of the codons have been translated, and the ribosome has reached one of the three stop codons, UAA, UAG or UGA. Then the ribosome releases the mRNA and moves away. See Fig. 5-3.

Several ribosomes can work on translating a piece of mRNA simultaneously. Once one ribosome has moved away from the start codon, a second ribosome can latch on and begin translating. This allows for several polypeptide chains to be generated from a single mRNA strand in a very short time. In some cases they can begin to take their active place in the protein they belong to even before they're complete.

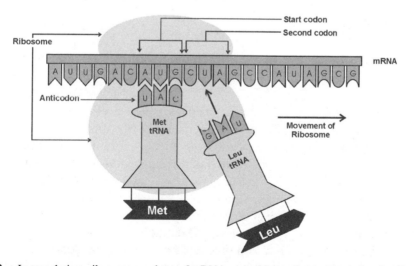

Fig. 5-3.   In translation, ribosomes and transfer RNA assemble a polypeptide chain of amino acids according to the blueprint copied by the mRNA from the cell's DNA.

However, some polypeptide chains don't become active until they've been modified. Some have to be *phosphorylated* (they need one or more phosphate groups added to them), some must be *glycosylated* (they need one or more carbohydrate groups added to them), and some need to be partially digested by *peptidase enzymes,* which break up protein into smaller units.

One important example is insulin. It's synthesized as a single-chain protein called *proinsulin.* It doesn't become active until peptidase enzymes make two internal cuts in the chain, removing 31 amino acids, and producing two polypeptide chains bound together.

How Genes Are Regulated

Remember that every cell in an organism contains the complete genetic code for the entire organism. That being the case, it's obvious that only certain genes within each cell are activated at any given time—specifically, those genes that either code for the proteins necessary to build the structures required by a daughter cell after cell division, or those genes that code for the proteins that the cell contributes to the operation of the organism as a whole.

Regulation of gene expression can (and does) occur at various points during the transcription and translation processes described above.

CROSS-REFERENCE

In this chapter, the focus is on the regulation of genes in eukaryotes. For more discussion on the process in prokaryotes, see Chapter 9.

As noted earlier, special DNA sequences called promoters indicate where transcription should begin. These promoters, as you'd expect, are usually located adjacent to the gene they promote. However, the promoters are stimulated by other sections of DNA called *enhancers,* which can be hundreds or even thousands of base pairs away in either direction. DNA binding proteins called *activators* or *repressors* (depending on their function) bind to the enhancer sequences, and interact with a series of other proteins that ultimately connect to the framework of proteins interacting at the promoter. The result is an increase or decrease in the number of RNA polymerase molecules transcribing the gene.

Promoters and enhancers together are called *cis-acting* elements, because they are located on the same DNA strand as the gene they control.

There are also *trans-acting factors,* which can coordinate the regulation of genes occurring on different DNA molecules in different chromosomes. Proteins called *transcription factors,* which are themselves coded by other distinct genes, bind to specific DNA sequences in promoters and either promote or repress transcription.

The activation and suppression of genes occurs in response to one or more signals or combination of signals, which can be either *exogenous* (coming from outside the organism) or *endogenous* (coming from other cells).

An example of an exogenous signal would be the absence of light. Plants start to lose their green color after several days in darkness because of the loss of enzymes that are the catalysts for synthesizing chlorophyll. When light is returned, so do the enzymes, and so does chlorophyll. A protein called phytochrome, which is bound to a light-absorbing pigment, is inactive in the dark; light activates it, and it's believed it then acts as a transcription factor for production of many of the enzymes crucial to photosynthesis.

An example of an endogenous signal would be a *hormone.* These substances, produced by one cell type, influence many other cell types. Although they're transported throughout the organism, only cells that have receptors for them on their surfaces are affected. The interaction of the receptor and the hormone produces a signal in the cell that results in transcription of particular genes.

Gene regulation can also occur during the processing of mRNA, through a mechanism called *alternative splicing.* In eukaryotes, every gene is made up of regions that carry genetic code (*exons*), interspersed with regions that don't (*introns*). Part of the process of transcription is thus the removal of introns and the splicing together of exons. (In prokaryotes—primarily bacteria—genes consist of a single contiguous thread of DNA.) Variations in what is removed and what is spliced together changes the mRNA and thus changes the final protein.

The expression of genes can also be regulated at the translational level. There are three known methods.

1. *Altering the stability of the mRNA.* The longer mRNA remains stable, the more quickly and efficiently the protein whose code it carries is produced.
2. *Controlling the start and speed of translation.* Translation may never take place, or it may be hindered or aided, usually by the presence or absence of a particular nutrient or metabolite (a product of metabolism).
3. *Modifying the protein after translation.* As mentioned earlier, some proteins have to be modified, in one of a variety of ways, before they become active.

The precise details of the expression and regulation of genes in eukaryotes continues to be the subject of intensive research.

From the Micro to the Macro

The ultimate result of all this transcription and translation is, of course, an organism with certain traits. The traits are determined by the cells and cell products made by the proteins synthesized in accordance with the instructions contained in the DNA in the cell nuclei.

The genetic information an organism carries within it is known as its *genotype;* the physical expression of that genetic information is the *phenotype.*

In eukaryotes, relatively few traits are controlled by a single gene (or pair of genes). Usually, several genes must work together to determine a particular trait. These traits are thus called *polygenic.*

One example is height. Your total height is determined by the sizes of various body parts—legs, torso, neck, head. Since the size of each of these parts is controlled by several genes, your overall height is controlled by multiple genes. Skin, hair, and eye color are also polygenic traits. Because of their complexity, instead of finding just a few easily classifiable colors of any of them, we get an impression of continuous gradation.

Some genes influence how other genes are expressed in the phenotype. Modifying genes may change the affect of other genes. For example, how much impact a certain dominant cataract gene has on a person's vision depends on the presence of a specific allele for a modifying gene.

Regulator or *homeotic* genes initiate or block the expression of other genes. For example, they play an important role in growth, initiating and controlling the development of body parts, beginning shortly after conception and continuing until adulthood.

Some genes do not affect the phenotype except in the presence of certain environmental factors. It's thought, for example, that the gene that causes multiple sclerosis may be activated by the Epstein-Barr virus.

On the other side of the scale from multiple genes that determine traits are single genes that determine multiple traits. This is called *pleiotropy.* Sickle-cell anemia, mentioned earlier, is one example. This defect is produced by a single gene, but the results include pain, poor development, and heart, lung, eye, and kidney problems. Albinism is another pleiotropic trait which affects eyesight as well as resulting in a lack of pigmentation in skin, hair, and eyes.

By reading the genetic code of humans and other organisms, researchers hope to link more and more genes to specific traits—particularly in humans, where troublesome traits like various genetic diseases and disorders plague millions of people.

In the next chapter, we'll take a look at the efforts to read the genetic code in its entirety—in humans and other organisms.

Quiz

1. Who first suggested DNA could be read as a code for producing the amino acids in a polypeptide chain?
   (a) James Watson
   (b) Francis Crick
   (c) Alexander Fleming
   (d) Jim N. E. Cricket

2. How many amino acids make up all life on earth?
   (a) 5
   (b) 10
   (c) 20
   (d) 200

3. What is each three-letter "word" of the genetic code called?
   (a) A codon.
   (b) An aminocon.
   (c) A proton.
   (d) A necronomicon.

4. The process of copying the genetic information from DNA to mRNA is called
   (a) transfixion.
   (b) transportation.
   (c) transliteration.
   (d) transcription.

5. The process of building chains of amino acids from the genetic information copied into mRNA is called
   (a) transmission.
   (b) transubstantiation.
   (c) translation.
   (d) transmutation.

6. What tiny structures in the cytoplasm of the cell control the process of building amino acid chains?
   (a) ribosomes
   (b) riboflavors
   (c) ribocops
   (d) ribozoids

7. Cis-acting elements are sections of a strand of DNA that activate or repress genes located
   (a) in other chromosomes.
   (b) in other cells.
   (c) on the same strand of DNA.
   (d) in the mitochondria.

8. The genetic information an organism carries is its genotype; the physical expression of that genetic information is its
   (a) phenotype.
   (b) linotype.
   (c) tintype.
   (d) typetype.

9. Physical traits determined by more than one gene are called
   (a) polyamorous.
   (b) polyphenic.
   (c) polyester.
   (d) polygenic.

10. Another name for regulator genes that initiate or block the action of other genes is
    (a) homologous.
    (b) homeotic.
    (c) homogenized.
    (d) harmonious.

Genomes—Reading the Genetic Code

Learning how to read the genetic code was one thing; actually reading it was another. The key to doing so was finally provided in 1977 by Fred Sanger of the U.K. Medical Research Council.

Sanger had won the Nobel Prize in Chemistry in 1958 for his discovery of the order of the 51 amino acid building blocks that make up the two chains of insulin. By following up his insulin breakthrough with a method for sequencing DNA that is still the basis for most of today's automated DNA sequencing technology, he became one of only four people to win two Nobel Prizes.

SEQUENCING DNA

Sanger's *chain termination* or *dideoxy* method of DNA sequencing begins with the synthesis of the single-strand DNA to be sequenced, altered by the addition of a radioactive marker that always shows up at the same end. Next, four different mixtures are set up, each one containing all the necessary nucleotides for

DNA replication but also containing a small amount of a modified nucleotide that can bond to the DNA strand—but then nothing can bond to it. In each mixture, a different one of these modified nucleotides (called a *dideoxynucleotide*) is included. Each dideoxynucleotide causes any DNA strand that incorporates it to terminate at a specific base.

In each of the four mixtures, the result is a mixture of strands of DNA of different lengths (the length determined by where—or if—they incorporated the dideoxynucleotide), every strand ending in the same base, and all marked with a radioactive tag.

Placing electrodes in a gel containing DNA causes the negatively charged DNA to migrate toward the positively charged electrode, or anode. This is called *electrophoresis*. The shortest pieces move faster and the longest pieces move slower. Since the DNA carries a radioactive tag, it can affect photographic film. Once the DNA has been spread out from shortest to longest fragments, the results can therefore be captured on film as a series of bands, each one marking a place in the DNA sequence where a particular base occurs.

The Maxam-Gilbert Method

At about the same time that Sanger was coming up with his DNA sequencing method, Walter Gilbert of Harvard University and graduate student Allan Maxam developed an alternate method, for which Gilbert went on to share Sanger's 1980 Nobel Prize.

Whereas Sanger's method relied on enzymes, the Maxam-Gilbert method used chemicals that acted on some bases more effectively than others, resulting in breaks in the DNA where those bases appeared.

Sanger's method, however, proved easier and more popular, and so has mostly eclipsed the Maxam-Gilbert method.

By comparing the bands generated by each of the four mixtures side by side, you can read the complete sequence of the DNA strand (see Fig. 6-1).

In July of 1984, just seven years after announcing his new DNA sequencing method, Sanger and his colleagues at the U.K. Medical Research Council completed the sequencing of the DNA contained in the Epstein-Barr virus. This is a virus of the herpes group that is the cause of infectious mononucleosis. The sequence was 5,375 bases long. It was not only the first complete genome of a DNA virus (or anything else) ever compiled—and the largest sequence of DNA

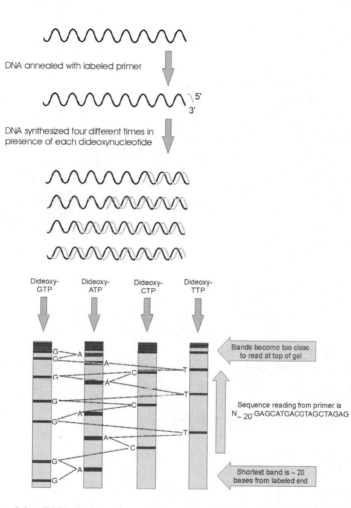

Fig. 6-1.   DNA can be sequenced by comparing the results of four reactions, each of which creates fragments ending in one of the four bases.

compiled to that time—it also demonstrated that genes could overlap on a strand of DNA, actually sharing sections of code.

It also showed just how daunting the task of unraveling the complete genetic code of any larger organism would be. If virus DNA was 5,375 bases long, how long would the DNA of a human being be?

The answer: more than three billion bases. Table 6-1 compares the genome sizes of several different organisms, from viruses to humans.

Table 6-1. Genome Sizes in Various Organisms

Genome	Group	Size (kb)	Number of Genes
Eukaryotic nucleus			
Saccharomyces cerevisiae	Yeast	13,500	6000
Caenorhabditis elegans	Nematode	100,000	13,500
Arabidopsis thaliana	Plant	120,000	25,000
Homo sapiens	Human	3,000,000	100,000
Prokaryote			
Escherichia coli	Bacterium	4700	4000
Haemophilus influenzae	Bacterium	1830	1703
Viruses			
T4	Bacterial virus	172	300
HCMV (herpes group)	Human virus	229	200

THE HUMAN GENOME PROJECT

In his book *Cracking the Genome: Inside the Race to Unlock Human DNA* (The Free Press, New York, 2001), Kevin Davies points out that if the sequence were written as a conventional book, each page would contain about 3000 characters, an average gene would run five pages, the DNA of an average chromosome would require about 200 books of about 300 pages each, and the entire human genome would require a library of 4000 books.

Yet it was precisely this daunting task that geneticists began talking about achieving. In fact, less than a year after the MRC scientists published the DNA sequence of the Epstein-Barr virus, a meeting was held at the University of California to discuss the feasibility of sequencing the human genome.

Throughout 1986, thanks to the efforts of a number of distinguished scientists and some exciting technological advances, the idea gained momentum. In 1987, an advisory panel suggested the U.S. Department of Energy (DOE) should spend $1 billion on mapping and sequencing the human genome over the next seven years. That same year, the first automated DNA sequencing machine went on the market (see sidebar).

The First Automated DNA Sequencer

Both the Sanger and Maxam-Gilbert methods of sequencing DNA were time-consuming, labor-intensive, and expensive. It seemed obvious that what was needed was some way to automate the procedure—and that's just what Leroy Hood provided geneticists in 1986, just as serious discussion about sequencing the human genome was getting underway.

Working with several others, Hood, a biologist at the California Institute of Technology, improved on Sanger's method. Sanger used radioactive labels, which posed several difficulties: They were unstable, they were unhealthy, and they required separate gels for each of the four DNA bases.

Hood developed a new method that used a different-colored fluorescent dye for each of the four DNA bases. With color coding, you no longer had to run separate reactions on separate gels.

The use of dyes also allowed Hood to automate the process of data gathering. In Sanger's original version of the technology, the sequence produced by the effect of the radioactive tags of the various lengths of DNA on photographic film produced an "autoradiograph" that a human reader then had to transfer to a computer.

In Hood's automated machine, a laser beam shone on the fragments of DNA caused the various fluorescent dyes to glow. That glow could be detected and fed as digital information directly into a computer.

Hood's machine was brought to market in June of 1986 by Applied Biosystems Inc., and has been constantly improved since. By 1999, a fully automated instrument could sequence up to 150 million base pairs annually. New variations are even faster.

In 1988, the DOE and the National Institutes of Health agreed to join forces on the Human Genome Project, with none other than James Watson as its director. After a considerable amount of discussion, wrangling, and planning, the U.S.'s public Human Genome Project was declared to have officially begun in October of 1990.

Watson resigned just two years later in a dispute over the patenting of partial genes by the National Institute of Health (NIH). He was replaced in 1993 by Francis Collins—the discoverer of the gene for cystic fibrosis.

Shortly before Collins took over, biologist J. Craig Venter left the NIH to set up The Institute for Genomic Research (TIGR), a non-profit company in Rockville,

Maryland. Its sister company Human Genome Sciences was set up to commercialize the products from TIGR. In 1998, Venter shocked the scientific world by setting up a new company called Celera and announcing that it would sequence the human genome within three years for just $300 million. Put simply, Venter believed he had a better way to sequence the genome.

Venter had long had an interest in speeding up the rate of discovery in genetic research. Shortly after the Human Genome Project got underway, he had demonstrated an efficient way to find genes and explore their functions. Called *Expressed Sequence Tags* (ESTs), they were based on the development of *complementary DNA* (cDNA). cDNA is DNA translated from mRNA molecules—the molecules created as genes are being read. cDNA carries the same information as mRNA, but has DNA's advantage of being far sturdier and living longer. ESTs are cloned segments of cDNA molecules that have been partially sequenced—typically to the extent of several hundred bases at both ends.

In 1991, Venter published the results of a pilot project using ESTs. He proved that ESTs could be used to identify new genes, could be used to help map known genes to chromosomes, and even demonstrated the accuracy of the new automated sequencing technology.

Venter followed that up in 1995 by publishing, with colleagues, the first completely sequenced genome of a self-replicating, free-living organism (in other words, something that wasn't a virus)—the bacteria *Haemophilus influenzae Rd*. Its genome was 10 times longer than that of any virus that had been sequenced, and Venter used an entirely new approach to sequence it. Called *whole-genome random sequencing* (but better known as *random shotgun sequencing*), it set out to assemble a complete genome from partly sequenced DNA fragments with the help of a computer model.

Copies of DNA from the bacteria were cut into pieces of random lengths between 1600 to 2000 base pairs. These bits were partly sequenced at both ends for several hundred base pairs. (A few longer fragments, 15,000 to 20,000 base pairs long, were also sequenced.) These sequences were fed into a computer, which compared, clustered, and matched them. Nonrepeating sequences were identified first, then repeating fragments. The longer fragments helped provide the correct order for some of the very repetitive, almost identical sequences. Various techniques were used for filling in gaps.

It only took a year for the genome of *Haemophilus influenzae Rd* to be sequenced: all 1,830,137 base pairs and 1749 genes identified from 24,304 DNA fragments. The success proved that random-shotgun sequencing could be applied to whole genomes quickly and accurately.

All of which meant that Venter's announcement that he was targeting the human genome and expected to complete it before the publicly funded project sent the publicly funded project into overdrive. In October of 1998, the NIH and DOE set a new goal of having a working draft of the human genome by 2001 (the same time Venter said he'd have one) and moved the completion date for the finished draft from 2005 to 2003. Just a few months later they moved the completion date of the rough draft to the spring of 2000.

And sure enough, in June of 2000, at a ceremony at the White House, the Human Genome Project and Celera jointly announced their rough drafts of the human genome sequence. HGP published highlights of its working draft (about 90 percent complete) in *Nature* in February of 2001, while Celera published highlights of its draft in *Science* that same month. Analysis of the draft genome initially indicated that humans have approximately 30,000 genes; continuing analysis has resulted in that being revised downward, to fewer than 25,000 genes.

The actual number, whatever it is, is certainly far smaller than the original estimates—some of which ran as high as 140,000 genes. There had long been a belief that one gene coded for one protein. But 25,000 genes aren't enough to account for all the different proteins in the human body. Instead, it appears that one gene can actually direct the synthesis of many different proteins—five or six on average per gene. (Mechanisms for achieving that include alternative splicing, mentioned in the last chapter.)

A "final draft" of the full human genome sequence was published by the Human Genome Project in April of 2003. However, since there are some elements of the genome that can't be sequenced with current technology, our knowledge of the human genome still isn't entirely complete.

Sequencing Other Species

Of course, the same sequencing technology used on humans can be used on other animals and plants, as Venter demonstrated. More than 200 organisms have had their genomes sequenced so far. Among them:

1. Yeast. In April of 1996, some 600 scientists from all over the world finished sequencing the genome of baker's yeast, *Saccharomyces cerevisiae*. That wasn't just in the interest of better bread: Yeast carries versions of many human genes and has been used in the laboratory for decades.

2. *Methanococcus jannaschii*. This microbe's genome was of particular interest because it's a member of the Archaea, thought to have evolved separately

from either eukaryotes or prokaryotes. It lives near hydrothermal vents on the sea bottom, in temperatures near boiling, and under enormous pressure. Scientists in Maryland sequenced its genome in 1996, and found many genes no one had seen in any other organism. Several more "extremophiles" have had their genomes sequenced since then. Among other things, the research has led to the discovery of natural proteins that can withstand high temperatures, and thus might have commercial applications.

3. *Caenorhabditis elegans.* This is a translucent, microscopic worm that has been used in genetics studies for years. It lives in the soil, grows to just a millimeter in length, and contains exactly 959 cells. Biologists like it because they can see inside it to closely observe biological development. It also shares many genes with humans.

4. *Drosophila melanogaster.* The genome of Drosophila, darling of geneticists since Columbia University's famous Fly Room, was sequenced in 1996 by Celera Genomics and the Berkeley Drosophila Genome Project (BDGP). The BDGP had about a fifth of the genome sequenced when Celera offered to complete the job—quickly and without federal funding—as proof of its capabilities and to check the accuracy of its "shotgun" method. Sequencing began in May of 1999 and was completed in September. The genome was assembled over the next four months. The fruit fly revealed 13,601 genes. More remarkably, a survey of 269 sequenced human genes whose mutations cause disease showed that 177 of them had a closely related gene in the *Drosophila* genome—proof of the research value of comparative genomics.

5. The mouse. Mice are used extensively in all kinds of laboratory research. In 2001, Celera Genomics announced the completion of a draft mouse genome sequence; in December of 2002 the international Mouse Genome Sequencing Consortium also completed a draft of the mouse genome. Comparison of the mouse genome and the human genome revealed that humans and mice have about 200 genomic blocks containing the same genes (albeit arranged on different chromosomes). Both species have about the same number of genes, and even the same "non-gene" or "junk DNA" regions. However, the mouse genome is about 14 percent smaller than the human genome.

6. The rat. Rats and mice are used in different ways in laboratory research, so scientists felt it was worthwhile sequencing the rat genome separately from the mouse genome. In 2004, a high-quality draft of the rat genome (about 90 percent complete) was unveiled by the Rat Genome Sequencing Project Consortium. They reported that rats have many of the genes whose mutation can cause disease in humans. Again, they found roughly the same

number of genes (between 25,000 and 30,000), although the rat genome is larger than the mouse genome (but smaller than the human genome).

7. **Chimpanzees.** The NHGRI announced a draft sequence of this closest of human relatives in 2003. Humans and chimpanzees have approximately 98 percent of their DNA in common. Some diseases, including AIDS, Alzheimer's, and malaria, affect the two species differently; comparing genomes may provide clues to those differences.

Among the other organisms whose genomes have been sequenced are flies, mosquitoes, dogs, poplar trees, pufferfish, two types of rice (a final, definitive rice genome-sequencing project is currently underway) and many, many different species of bacteria, from notorious disease causers (such as *Yersinia pestis,* which causes plague) to the potentially useful (such as *Pseudomonas putida,* a soil bacteria that could be used to cleanup pollutants.) Scientists continue to work on sequencing the genomes of additional organisms.

Genome Mapping

A map is a graphical representation that helps you figure out where you are and get to where you want to go. A genome map, therefore, is a graphical representation that helps researchers figure out where things are in the genome and where to go to find other things that might be of interest or importance.

A genome map can contain all sorts of information—the location on the genome of specific genes or regulatory sites—but it usually also contains large gaps, because a genome map is constantly under revision as scientists learn more about a particular genome.

In its simplest form, a genome map is a straight line, just like the DNA molecules that make up the genome. Along its length are the various landmarks, identified with a series of letters and numbers that identify the various features to researchers.

A map is not the same thing as a genome sequence. The location of some genes on the map can be figured out without sequencing. In fact, a genome map can help you sequence a genome by providing clues as to where specific pieces of DNA belong in the overall genomic jigsaw puzzle.

But not only that, a map provides valuable information that the genome sequence doesn't. The genome sequence is just that, a sequence of the same four letters in endless but mind-numbing variation. Even a scientist can't look at a

DNA sequence and immediately figure out what it does. Slotting that sequence into its correct position in a map provides valuable clues as to what purpose— if any—it serves (see Fig. 6-2).

Here's one way researchers might use a genome map. Suppose they want to pinpoint the location of a particular disease-causing gene. They would first study several families afflicted by the disease to see if inheritance of the disease is tied into any other genetic landmarks. Anything else that tends to be inherited along with the disease is likely to be located fairly close to the disease-causing gene on the same chromosome, and can become a marker for the gene being sought.

By identifying several markers whose location on the chromosome are known, scientists can get a pretty good approximation of the location of the disease-causing gene—within a few million base pairs, anyway. Then they can focus in on that part of the genome and look for a gene that has a different sequence in disease-affected people than in healthy people, or for a gene whose function might be related to the disease in some way.

This is precisely how the genes for cystic fibrosis and Huntington's disease were identified. However, it's laborious and time-consuming, so the goal continues to be developing more detailed genome maps—maps that can someday take researchers precisely to the location on the genome in which they're interested.

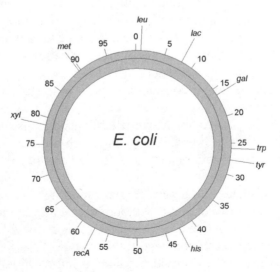

Fig. 6-2. A typical genome map.

The two main types of genome maps are *genetic-linkage maps* and *physical maps.*

Genetic-linkage maps show the order of genes on a chromosome and the relative distances between them. This is the kind of map produced by A. H. Sturtevant, which we learned about back in Chapter 4.

Sturtevant's map had to be based on traits that were physically visible in the fruit flies with which he worked. Today, much more complex genetic linkage maps can be made by tracing the inheritance of specific DNA sequences.

Physical maps indicate the number of DNA base pairs between one landmark and the next. They're based on sequence-tagged sites, or STSs. An STS is a sequence of DNA that is found in only one place in the genome, and is a few hundred base pairs long. It can be part of a gene, but it doesn't have to be.

Since an STS occurs in only one location, any time an STS shows up in a fragment of DNA during sequencing, it tells you where that fragment belongs in the genome.

New genomic maps are now being drawn up that combine features of both types of maps. Maps that include the sequence and location of all of an organism's genes now exist for more than 150 organisms. However, most of them are viruses with very small genomes, an indication of the difficulties genomic cartographers face.

Genome Variation

As noted earlier, more than 99.9 percent of your DNA sequence is exactly the same as that of your next-door neighbor or any other human being in the world. In other words, your genome differs from mine about 1 in every 1200 to 1500 DNA bases. But that still leaves lots of room for variation in a genome consisting of almost 3 billion base pairs.

These differences arise because of mutations—wrong bases incorporated into the daughter strands of a DNA molecule as it replicated somewhere along the way. Mutations in a sex cell can be passed along to the next generation. Each individual contains about 100 new mutations that arose in just this way.

Some parts of the genome are more likely to vary between individuals than others. As well, most of the variations occur in the "junk DNA" outside of genes, and so doesn't have any physical affect.

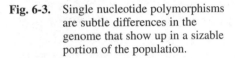
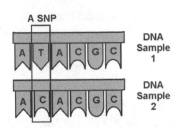

Fig. 6-3. Single nucleotide polymorphisms are subtle differences in the genome that show up in a sizable portion of the population.

Technically, genetic variations are defined as either *mutations* or *polymorphisms.* If a possible sequence shows up in less than one percent of people, the variation is called a mutation. If its shows up in one percent or more, it's called a polymorphism.

About 90 percent of the variation in the human genome shows up as single nucleotide polymorphisms, or SNPs ("snips"), involving only one base—a T instead of a C, for instance (see Fig. 6-3). Since most of the genetic variation in the genome comes from these tiny changes, scientists are beginning to believe that many of the physical differences that make each individual unique comes from these tiny variations. SNPs may also help to explain differing susceptibilities to various diseases.

In the next chapter we'll take a closer look at mutations and other forms of genetic variation, and their impact.

Quiz

1. Fred Sanger invented a method for
    (a) replicating DNA.
    (b) dissolving DNA.
    (c) sequencing DNA.
    (d) identifying DNA.

2. Passing an electric current through a gel in order to sort DNA fragments by length is called
    (a) electrophoresis.
    (b) electrolysis.
    (c) electroplating.
    (d) electromagnetism.

3. Approximately how many base pairs does the human genome contain?
    (a) three thousand
    (b) three million
    (c) three billion
    (d) three trillion

4. What did the first automated sequencing machine use instead of radioactive molecules to tag DNA fragments?
    (a) left-handed sugars
    (b) RNA
    (c) heavy metals
    (d) fluorescent dyes

5. What was the first organism whose genome was sequenced using J. Craig Venter's "shotgun" approach?
    (a) The *Haemophilus influenzae* bacterium.
    (b) The *E. coli* bacterium.
    (c) The malaria mosquito.
    (d) The mouse.

6. Which of these numbers is closest to the current estimate of the number of genes in the human genome?
    (a) 2500
    (b) 25,000
    (c) 250,000
    (d) 2,500,000

7. The genomes of what two rodents have recently been sequenced?
    (a) Mice and rats.
    (b) Beavers and gophers.
    (c) Squirrels and chipmunks.
    (d) Hamsters and gerbils.

8. The two main types of genome maps are
   (a) circular and linear.
   (b) chromosomal and whole-genomic.
   (c) genetic-linkage and physical.
   (d) Rand-McNally and Fodor's.

9. One individual's genome differs from another individuals at the rate of
   (a) about 1 in every 1200 bases.
   (b) about 1 in every 12,000 bases.
   (c) about 1 in every 120,000 bases.
   (d) about 1 in every 1,200,000 bases.

10. A polymorphism is a genetic variation that shows up in
    (a) ten percent or more of the population.
    (b) five percent or more of the population.
    (c) one percent or more of the population.
    (d) less than one percent of the population.

Mutations— Misreading the Code

One of the members of the research team at Columbia University headed by Thomas Hunt Morgan was a young man in his 20s named Hermann J. Muller. The basis of all of the Morgan team's work with fruit flies was mutations. It was the appearance of previously unseen traits in the flies—which could then be bred into subsequent generations—that allowed them to map specific genes to specific chromosomes.

Unlike most early geneticists, Muller had a strong interest in the physical and chemical makeup and operation of genes, and in late 1926 (by which time he was at the University of Texas), he began a series of experiments designed to reveal if high levels of radioactivity could induce mutations in fruit flies.

Muller exposed male flies to radiation, then mated them with non-irradiated female fruit flies. His experiments were a success. In just a few weeks he managed to induce more than 100 mutations in the offspring of the irradiated flies, half as many mutations as had been discovered to naturally occur in *Drosophila* fruit flies over the previous 15 years of research. The mutations ranged from the minor to the deadly.

Muller suggested that the particles of radiation, as they passed through the chromosomes of the male fruit flies, randomly affected the molecular structure of individual genes. In some cases that destroyed the genes; in others, it altered their chemical functions.

Since then, researchers have discovered many more artificial *mutagens*, as anything that causes mutations is called, and have also learned a great deal more about the many ways that the genetic code can be altered.

Types of Mutations

It's likely that the fruit flies Muller observed suffered an even higher rate of mutation than he realized. By definition, a mutation is any change in the sequence of DNA. That means a mutation can occur anywhere in the genome. And since most of the genome is made up of "junk DNA" that serves no apparent purpose, most mutations go unnoticed.

A mutation only alters the physical characteristics of an organism if it occurs inside the DNA sequence of a gene (see Fig. 7-1 for an example).

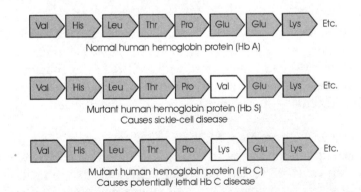

Normal human hemoglobin protein (Hb A)

Mutant human hemoglobin protein (Hb S)
Causes sickle-cell disease

Mutant human hemoglobin protein (Hb C)
Causes potentially lethal Hb C disease

Fig. 7-1. These three amino acid sequences show how a tiny change in the gene for hemoglobin can produce major effects. The beginning of one of the amino-acid chains in the normal protein is shown at the top. Underneath it is the amino acid chain for an abnormal version of the hemoglobin protein: Valine has substituted for glutamic acid in the sixth position. This single change, which results when the codon GAA is mutated to GUA, produces sickle-cell disease, the effects of which can range from a mild anemia (if the individual still has one good copy of the mutated gene) to a lethal one (if the individual has two copies of the mutated gene).

Although Muller generated mutations in fruit flies by exposing them to high levels of radiation, mutations occur all the time in organisms, sometimes just through errors in normal cellular operations, sometimes through the effects of the environment. These spontaneous mutations occur at a rate characteristic of a specific organism, which is sometimes called the *background level.*

The most common type of mutation is the *point mutation*, which changes only a single base pair in the normal sequence. It can be caused in one of two ways:

1. The DNA is chemically modified so that one base changes into a different base.
2. DNA replication malfunctions, causing the wrong base to be inserted during DNA synthesis.

Whatever their cause, point mutations can be divided into two types:

1. *Transition.* This is the most common type. In a transition mutation, one pyrimidine is replaced by the other, or one purine is replaced by the other: For example, a G-C pair becomes an A-T pair, or vice versa.
2. *Transversion.* In this less common type, a purine is replaced by a pyrimidine, or vice versa: For example, an A-T pair becomes a T-A or C-G pair.

Nitrous acid is a mutagen that can bring about a transition mutation. It converts cytosine into uracil. Cytosine would normally pair with G, but uracil pairs with A. As a result, the C-G pair becomes a T-A pair when the A pairs with T in the next replication cycle. Nitrous acid can have the same affect on adenine, turning an A-T pair into a C-G pair.

Another cause of transition mutations is *base mispairing*. These occur when something introduces an abnormal base into the DNA strand, which then pairs with an unusual partner instead of the one that would normally be expected. Thus, the base pair is altered completely during the next replication cycle.

The effect of these kinds of mutations depends on where in the sequence the transition takes place. Since a change in a single base pair alters only one codon, and hence only one amino acid, the resulting protein—though it may be damaged—may still be able to carry out some of its normal activity.

Far more damaging than point mutations are *frameshift mutations*. Remember that the genetic sequence is read in non-overlapping triplets. That means that there are three ways to translate any nucleotide sequence, depending on where you start. If a mutation inserts or deletes a single base, then it causes a frameshift and the entire subsequent sequence of the gene will be misread. That means the entire amino acid sequence is changed, and most likely the protein that is supposed to be produced is completely nonfunctional.

Frameshift mutations are caused by *acridines*, which are compounds that bind to DNA and alter its structure to the point that bases can be added or omitted during replication. The effect of mutation on the reading of the genetic sequence depends on where the insertions and deletions take place, and how they relate to each other (see Fig. 7-2).

Still another form of mutation is the insertion of long stretches of additional material into the genetic code. These insertions come from *transposable elements*, or *transposons*, sequences of DNA that can move from one site to another. First discovered in maize by geneticist Barbara McClintock in the 1950s, transposons are short DNA elements that can jump to a new spot in the genome (which is why you'll sometimes hear them referred to as "jumping genes.") Sometimes they take adjacent DNA sequences with them. They generally include one or more genes, one of which is for an enzyme called *tranposase* that they require for movement within the cell.

There are also *retrotransposons* that can't move themselves. Instead, they use mRNA, which is copied to DNA, which then is inserted in a new spot in the genome. Retrotransposons are related to retroviruses.

If a transposon is inserted into a gene, it messes up the coding sequence and the gene will most likely be switched off. Transposons may also contain signals to terminate transcription or translation, which effectively blocks the expression of other genes downstream. This kind of effect is called a *polar mutation*.

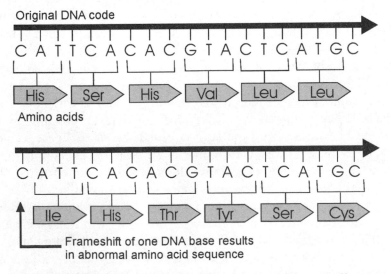

Fig. 7-2. Some of the ways in which a frameshift mutation can affect the reading of the genetic sequence.

Retrotransposons are well represented in mammalian genomes. In fact, as much as 40 percent of the genome may be made up of these sequences, which is one reason the genome contains so much "junk" DNA. Retrotransposons can be *SINEs* (for Short Interspersed Elements), just a few hundred base pairs long, or *LINEs* (Long Interspersed Elements) 3000 to 8000 base pairs long. For example, the human genome contains about 300,000 members of one SINE that appears to serve no function and may be nothing more than *selfish DNA* that exists only to replicate itself.

Unlike point mutations, mutations resulting from transposable elements can't be artificially induced by mutagens.

Mutations aren't necessarily irreversible. A point mutation may revert to its original sequence either through the restoration of the original sequence or through a mutation somewhere else in the gene that compensates for it.

An insertion of additional material can, obviously, revert through the deletion of the inserted material.

A deletion of part of a gene, however, can't revert.

Mutations may also occur in other genes as a way of "routing around" the damage caused by the initial mutation, resulting in a double mutant that still has a normal or near-normal phenotype. This is called *suppression*, and it comes in two forms, *intergenic* and *intragenic*.

An *intergenic suppressor mutation* suppresses the effect of a mutation in another gene, sometimes by changing the physiological conditions so the protein coded for by the suppressed mutant can still function. Sometimes this actually changes the amino acid sequence of the mutant protein.

An *intragenic suppressor mutation* suppresses the effect of a mutation in the same gene in which it is located, sometimes restoring the correct reading frame after a frameshift, sometimes producing new amino acids at different sites in the sequence that compensate for the amino acids changed by the first mutation. This is also called a *second-site reversion*.

Not all base pairs in a gene are as susceptible to mutations as others. Instead, mutations tend to cluster around hot spots in the gene sequence, places where the incidence of mutation is 10 or even 100 times higher than would be expected if the location of mutations was purely random. The location of these hot spots varies by type of mutation and by mutagen.

In *E. coli* bacteria, for example, hot spots appear in the DNA where a modified base called 5-methylcytosine occurs. This base occasionally undergoes a *tautomeric shift*, a rearrangement of a hydrogen atom. This results in G pairing with T instead of C, and after replication produces one wild-type G-C pair and one mutant A-T pair. (In genetics-speak, *wild-type* DNA sequences are those that occur most often in nature.)

Many mutations produce no apparent affect; they're called *silent mutations*. Sometimes a mutation is silent because the changes don't affect amino acid production. Sometimes they're silent because, even though the mutation changed an amino acid, whatever amino acid replaced it in the protein doesn't change the protein's function. That's called a *neutral substitution.*

A mutation that inactivates or alters the function of a gene is called a *forward mutation*. A mutation that reactivates or repairs that gene, either by reversing the original mutation or working around it (as in the second-site reversion mentioned earlier), is called a *back mutation.*

As you can see, there are many different ways to categorize mutations, and a single mutation may belong to more than one category. Table 7-1 may clarify any confusion.

Table 7-1. Classifying Mutations

Size	
Point mutation	A change in a very small segment of DNA, i.e., a single base or base pair
Gross mutation	Any change involving more than just a few nucleotides, right up to the level of entire chromosomes
Effect on codon	
Silent	A change that does not alter the amino acid that is being coded
Nonsense	A change that alters a codon from one that codes for an amino acid to a stop codon that brings translation to a premature halt
Missense	A change to a codon so that it now codes for a different amino acid than it did originally; the new amino acid alters the function of the protein
Neutral	A change to a codon so that it now codes for a different amino acid than it did originally; however, the new amino acid behaves similarly to the original one and thus the protein continues to function
Frameshift	A shift in the reading frame caused by the addition or deletion of one or more nucleotides that creates a whole string of nonsense and missense codons downstream from the initial mutation
Effect on gene function	
Loss-of-function mutation	A mutation that causes a gene to stop functioning; if it's hereditable, it tends to be recessive
Gain-of-function mutation	A mutation that causes a gene to have a new or different function; if it's hereditable, it tends to be dominant

Table 7-1. Classifying Mutations *(Continued)*

Effects on DNA	
Structural mutations	Changes in the nucleotide content of the gene: these include substitution of one nucleotide for another, deletion of some portion of DNA, or insertion of some portion of DNA
Chromosomal rearrangements	Large structural changes caused by changing the location of a piece of DNA within the genome: these include the translocation of DNA to a different chromosome and a rotation or flip of the DNA sequence within the same chromosome
Origin	
Spontaneous	Mutations that occur during normal cellular activity
Induced	Mutations brought about by a mutagenic agent or something in the environment
Mutator	Mutations caused by mutations in other parts of the genetic sequence
Effect on the phenotype	
Subvital	Mutation reduces viability by less than 10 percent compared to the wild type
Semilethal	Mutation causes more than 90 percent but less than 100 percent mortality
Lethal	All individuals with this mutation die before reaching adulthood
Direction	
Forward	A mutation that alters the wild-type phenotype
Reverse or back	A mutation that changes an altered sequence back to its original sequence
Suppressor	A change from abnormal back to wild type, either through a mutation in the same gene but at a different site that results in a return to wild-type function (intragenic) or through a mutation in another gene that restores the wild-type function lost due to the original mutation (intergenic)
Cell type	
Somatic	Occurs in body cells, excluding sex cells, usually altering the phenotype only of the individual organism affected
Gametic	Occurs in the sex cells, producing a change that can be passed on to subsequent generations

Causes of Mutations

There are many causes of mutations. Some are simply spontaneous, arising through occasional errors in DNA replication and repair. Others, however, are induced—brought about by a mutagen or something in the environment. A substance or environment is said to be mutagenic if it consistently causes mutations at a rate higher than the background level.

There are three main types of mutagenic agents:

1. *Ionizing radiation.* Alpha, beta, gamma, and x-rays can all disrupt the normal DNA sequence, primarily by knocking out base pairs.
2. *Non-ionizing radiation.* Ultraviolet light can cause adjacent thymines on a DNA strand to bond together, which blocks DNA replication, requiring repair work. If the repair isn't completely successful, point mutations may arise.
3. *Chemicals.* Many different substances can interact with DNA in such a way that base-pair sequences are altered. There are three main types of chemical mutagens:
   a. *Base analogs.* These are chemicals that are so structurally similar to the bases in DNA that they can be incorporated into a growing DNA strand. But once there, they foul up proceedings by binding with bases other than the one with which the base they replaced would bind. Bromouracil is one example. It's structurally similar to thymine, which means it can be incorporated into a DNA strand in a T position, but it pairs more readily with G than A, which subsequently creates a strand with a G-C pair instead of an A-T one.
   b. *Base modifiers.* Some chemicals can change existing bases, and in so doing cause them to pair with bases other than the ones with which they would normally pair.
   c. *Intercalcalating agents.* These chemicals insert themselves directly into the DNA helix, interrupting replication and transcription. The result is usually insertions or deletions in the genetic code.

Most mutations are *somatic cell mutations*, which means they occur in the non-sex cells of the body. If they affect the phenotype in some way, it's for that individual only.

Occasionally, however, a mutation occurs in a germ cell. These gametic mutations produce a change that can be inherited by the offspring.

And at the end of the book, in Chapter 14, we'll talk more about gametic mutations as we consider the genetics-driven change in species we call evolution. But next, we'll take a look at one of the more dire results of somatic cell mutations: cancer.

Quiz

1. In an attempt to induce mutations, Hermann J. Muller exposed fruit flies to
   (a) chlorine.
   (b) acid fumes.
   (c) tobacco smoke.
   (d) radiation.

2. The term for something that induces mutations is
   (a) mutational multiplier.
   (b) mutator.
   (c) mutagen.
   (d) genesplicer.

3. What do we call a mutation that changes a single base pair?
   (a) a point mutation
   (b) a pair mutation
   (c) a base deletion
   (d) a nucleotide nullification

4. A mutation that removes a base pair and thus offsets the reading frame of the genetic sequence by one letter is called
   (a) a basal disruption.
   (b) a frameshift mutation.
   (c) a bump-and-run mutation.
   (d) a sliding sequence mutation.

5. Stretches of DNA that can move from place to place in the genome are called
   (a) wandering sequences.
   (b) genomic gypsies.
   (c) transposons.
   (d) quarks.

6. DNA that serves no function except to reproduce itself is called
   (a) greedy DNA.
   (b) selfish DNA.
   (c) rowdy DNA.
   (d) self-indulgent DNA.

7. A suppressor mutation is
   (a) a mutation that cancels out or works around the change produced by a previous mutation.
   (b) a mutation that prevents any other mutations from ever happening.
   (c) a mutation that suppresses the desire for food.
   (d) a mutation that results in serious mental depression.

8. Wild-type versions of genes are
   (a) those that cause organisms to act wildly.
   (b) those found only in wild animals, as opposed to domestic ones.
   (c) those that most often occur naturally.
   (d) mutated versions of normal genes.

9. Ionizing radiation, ultraviolet radiation, and some chemicals are all
   (a) causes of mutations.
   (b) things you're likely to find in a genetics lab.
   (c) used in cooking.
   (d) fun for young and old.

10. Gametic mutations are those that occur
    (a) in body cells, excluding the sex cells.
    (b) in sex cells.
    (c) in skin cells.
    (d) in eye cells.

Cancer—Genetics Gone Awry

In 1909, a farmer brought a chicken to the Rockefeller Institute for Medical Research in New York City. The farmer wasn't confused about the difference between human and veterinary medicine: He was there to see Francis Peyton Rous, a 30-year-old pathologist.

The farmer pointed out a cancerous tumor on the chicken and told Rous that other chickens in his flock also had tumors. Unlike "normal" cancer, this kind of cancer seemed to be contagious.

That intrigued Rous. He broke up cells from some of the chicken tumors and forced the resulting mixture through filters too small for even the tiniest bacteria to pass. When he injected the filtered material into healthy hens, they developed precisely the same kind of tumors as those from which the donors suffered. Not only that, but injecting the material into fertilized eggs also resulted in cancerous chickens.

At that time, no one had ever seen a virus. "Virus" was just a term for theoretical disease-causing organisms smaller than bacteria—but Rous concluded that viruses could cause cancer. The idea was radical then, but in 1966, Rous's work won him the Nobel Prize in Physiology or Medicine—at the age of 87.

In the interim, the science of genetics had caught up with Rous's pioneering work, and his work had, in turn, helped inform decades of research into the genetic basis of cancer.

What Is Cancer?

Cancer is a genetic disease in which cells multiply uncontrollably. They may form a localized *tumor,* a mass of identical cells, or they may spread throughout the body, causing tumors to grow in various locations. As the cancer cells multiply, they crowd normal cells, pushing against them or damaging them. This results in pain, damage to tissues and organs, and eventual death.

We now know that cancer is the result of multiple mutations that alter normal cells so that they undergo three major changes (see Fig. 8-1):

1. *Immortalization.* Most cells divide only a certain number of times—in culture, typically around 50 times—then stop and eventually die. Cancer cells can keep dividing indefinitely.
2. *Transformation.* Cancer cells do not observe the normal constraints on cell growth and division.
3. *Metastasis.* This is one of the most damaging features of cancer cells: They gain the ability to move away from their place of origin and invade other tissues.

Not all tumors are cancerous: Warts are tumors, but they aren't cancer. Tumors can occur in both plants and animals, but cancer occurs only in animals. That's because the thick cell walls of plants prevent metastasis from taking place.

The first step toward cancer is a *neoplasm*, a population of cells growing out of control. However, a neoplasm is considered benign unless it is capable of metastasizing, in which case it's considered malignant—and life-threatening.

The process by which a normal cell becomes a cancer cell is called *oncogenesis,* and *oncology* is the study of cancer, from which you can rightly gather that anything with the prefix onco- has something to do with cancer. (It comes from the Greek word *onkos*, which means bulk or mass—i.e., tumor.)

The Rous sarcoma virus, as the virus Rous discovered in his sick chickens became known, led directly to the genetic understanding of cancer in the early 1970s.

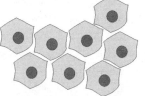

Primary Cells

Culture divides several times.

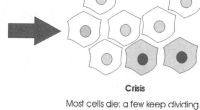

Crisis

Most cells die; a few keep dividing.

Transformation

Cells independent of anchorage or serum, no longer inhibited by contact, change shape, round up, form focus.

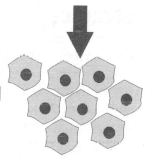

Immortalization

Divides indefinitely, but cells continue to adhere to whatever they're growing in, require serum, and are inhibited by contact.

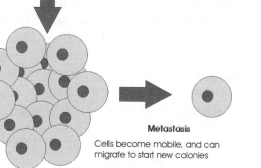

Metastasis

Cells become mobile, and can migrate to start new colonies

Fig. 8-1.   The process of turning normal cells into immortal, transformed, metastasizing cells can be carried out *in vitro* in the laboratory, as this diagram illustrates.

In 1970, David Baltimore of the Whitehead Institute for Biomedical Research in Cambridge, Massachusetts, discovered that cancer-causing viruses (others besides the Rous sarcoma virus had been discovered by that time), had genetic material made of RNA rather than DNA. They also produced an enzyme called *reverse transcriptase.* When these viruses injected their RNA into cells, reverse transcriptase turned the usual way of doing things upside-down. Instead of the cell's DNA being transcribed into RNA, the virus's RNA was transcribed into DNA, which was then inserted into the cell's genes. The virus's goal, like all viruses, was to hijack the cell's machinery and turn it into a viral factory. In the process, it sometimes also turned them cancerous.

CROSS-REFERENCE

For a more detailed look at the genetics of viruses, see Chapter 11.

But shortly thereafter researchers discovered a form of the Rous sarcoma virus that *didn't* cause cancer. They found that it lacked a large gene at the end of its genome—a gene the cancer-causing form had. They named the gene *src,* for sarcoma. Other cancer-causing genes were soon isolated in other cancer-causing viruses.

These *oncogenes* were thought to have originated with the viruses that delivered them into the cells they infected. But in the mid-1970s research led by Harold Varmus and Michael Bishop at the University of California at San Francisco revealed that just the opposite was true.

Using reverse transcriptase, their team made copies of the *src* gene from the Rous sarcoma virus, labeled them with a radioactive tracer, then mixed them with normal chicken DNA. To their surprise, the researchers found that *src* was already present in the normal DNA—and it was already active, even though the cells weren't cancerous. The *src* gene, or something very much like it, hadn't originated with the Rous sarcoma virus at all. Instead, it appeared much more likely that the virus had picked up *src* inadvertently at some point during its long interaction with chickens. Somehow, though, that normal chicken gene became a cancer-causer when the viruses re-injected it.

The researchers went on to discover versions of *src* in other fish, birds, mammals, and humans. The *src* gene obviously did something very important. The fact that it sometimes caused cancer was just an unfortunate side effect.

That work—in many ways a continuation of Rous's much earlier research—finally confirmed the speculation that cancer was indeed a genetic disease.

Since then, many other oncogenes have been discovered to be altered forms of normal cell genes. Today, the noncancer-causing versions of oncogenes are known as *protooncogenes.*

How Oncogenes Cause Cancer

Like every other gene, a protooncogene provides the instructions for constructing a particular form of protein. Around 1980, scientists discovered that the proto-oncogene form of *src* made the protein *kinase,* which adds phosphate groups to certain amino acids, a process called *phosphorylation.*

Phosphorylation is a normal part of cell activity. In fact, it's essential to some activities, especially cell growth. Normally, however, kinases only kick in when signaled to do so by other cell chemicals. Kinase-producing oncogenes produce kinases constantly.

Other protooncogenes carry out similar important tasks, all of which can be subverted in a way that promotes cancer. Some make growth factors, which make cells grow and reproduce; others make receptors, proteins to which growth-promoting substances attach, and which in turn send signals that activate the production of additional growth-related chemicals. The oncogene versions of all of these keep promoting growth when they should be shutting down.

As the much simplified diagram in Fig. 8-2 shows, cell growth is triggered by a complex process involving many different chemical signals. Protooncogenes are involved at every step of the path.

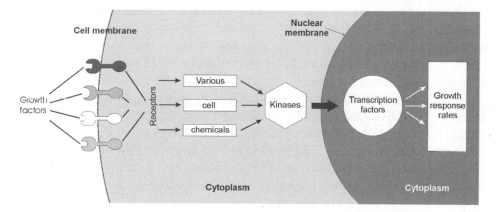

Fig. 8-2.   Cell growth begins with a signal from growth factors—chemicals sent to the cell from other cells. The growth factors attach to receptors in the cell membrane, which promote chemicals in the cytoplasm, which activate kinases, which activate chemicals that can pass through the wall of the cell nucleus to turn on transcription factors, which then turn on the genes that make the cell divide. Changes in any of the proteins involved in this complex cascade may trigger uncontrolled growth.

It doesn't necessarily take a large mutation to turn a protooncogene into an oncogene. In 1981, Robert Weinberg of the Whitehead Institute isolated the first oncogene from a human bladder tumor: *ras*. Weinberg and his coworkers found that *ras* differed from the protooncogene version of the same gene by only one base pair.

Some oncogenes become cancer causing when they move from one chromosome to another, or change locations on their normal chromosome, probably because they end up next to a gene that activates them.

Another Mutation That Causes Cancer

The conversion of protooncogenes to oncogenes is an example of a gain-of-function mutation: A normal gene begins functioning in a way it should not, and runaway cell growth ensues.

But a loss-of-function mutation can also cause cancer, as scientists discovered in the early 1980s. Just as there are genes whose normal function is to promote cell growth, there are genes whose normal function is to stop cell growth. If these *tumor suppressor genes* are switched off, uncontrolled cell growth can ensue just as surely as it can under the influence of growth-promoting oncogenes.

The first tumor suppressor gene to be identified was found in a rare type of cancer called *retinoblastoma* that affects the eyes of young children. This disease usually requires the life-saving removal of one or both eyes. What puzzled doctors about the disease was that it seemed to be inherited in some cases, but to arise spontaneously in others.

In 1971, Alfred G. Knudson proposed that children who develop the inherited form of retinoblastoma must have inherited two defective copies of a gene. However, he suggested, the cancer could also arise in children who inherited just one faulty copy of the gene and lost the second copy through random mutation, possibly during the period before birth when eye cells multiply rapidly. In rarer cases, a child who inherited normal genes but suffered mutations that made both of them inactive could also develop the disease. In other words, it always had the same ultimate genetic cause—two inactive genes—but how those genes became inactive could vary (see Fig. 8-3).

Next, Jorge Yunis of the University of Minnesota Medical School found that a part of chromosome 13 was missing in all cells of children with inherited retinoblastoma, but only missing in the tumor cells of children with the noninherited form of the disease.

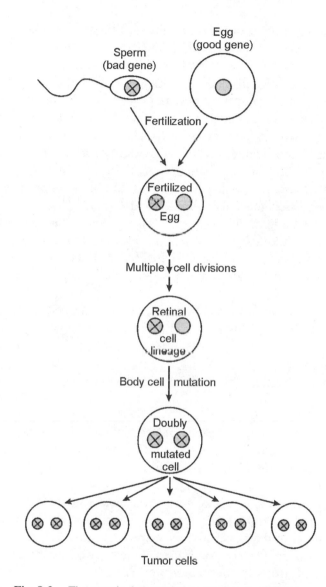

Fig. 8-3. The genetic damage that results in retinoblastoma is sometimes inherited is sometimes not. In this example, a child inherits one damaged gene from her father. In the (fairly likely) event that a mutation during eye development damages the healthy gene inherited from her mother, the mutated cell will no longer be able to control its growth. It will multiply and form a tumor. A tumor could also result even if the child inherited two normal copies of the gene, if both of them mutated before birth.

Many research groups set out to isolate the specific gene responsible. In the end, much of the work was carried out by Theodore Dryja, a doctor at the Massachusetts Eye and Ear Hospital. Although not formally trained in molecular biology, he began cloning pieces of chromosome 13 and comparing those pieces to DNA taken from both normal eye cells and retinoblastoma cells. In 1986, he located a DNA sequence that normal cells had but retinoblastoma cells did not have. He turned the samples over to the better-equipped lab run by Robert Weinberg, where Stephen H. Friend soon pinpointed the gene, dubbed *Rb* for retinoblastoma.

The *Rb* gene has since been found in other tissues besides the eye and linked to other forms of cancer. But it isn't the only tumor suppressor gene. Several other tumor suppressor genes have been found.

One of the most important—and common to a variety of cancers (including colon, bladder and breast cancer)—is *p53*. It's estimated to show up in as many as half of all the cancers that occur in the United States. It takes two defective copies of *Rb* to deactivate a cell's growth-suppression ability. It only takes one defective copy of *p53*.

That doesn't necessarily mean an instant case of cancer, though. Usually, the development of cancer requires more than one change in genes.

Carcinogens

Very few cancers are strictly inherited. Instead, an individual may inherit a gene that gives him an increased risk of cancer, but the cancer won't take hold until and unless cells suffer an additional mutation, or two or three. Ultimately, environment is more important than heredity.

Carcinogens are mutagens that turn cells cancerous. (All carcinogens are mutagens, but not all mutagens are carcinogens.) Examples include various chemicals, ionizing radiations, and, as mentioned already (and mentioned in more detail in Chapter 11), certain viruses. At least two major steps are required for their nefarious efforts to succeed, however:

1. *Initiation*. This is an initial mutation that results from a cell's first exposure to a carcinogen. It primes the cell to become cancerous, but doesn't by itself have that result.
2. *Promotion*. This involves one or more exposures to the same carcinogen or to different carcinogens (or even some substances not otherwise thought of as carcinogens) called *promoters*. Promoters eventually bring about whatever other mutations are necessary to turn cells cancerous. It's a gradual process that can take weeks in laboratory rats and years in humans.

So, the first mutation of a cell must be followed by subsequent mutations and other changes before cancer develops. Colon cancer, for instance, only develops after the activation of one oncogene and the inactivation of three growth suppressor genes, including *p53*. That's why cancer takes years to develop and is more common in older people.

Some carcinogens have to be altered chemically inside the exposed individual before they become active. This *metabolic activation* is usually brought about by enzymes in various tissues, especially the liver.

One reason certain substances cause cancer in some species and not in others is that some species may not have the enzymes required to turn certain inactive *precarcinogens* into active carcinogens.

The Smoking Gun of Smoking

Although the link between smoking and lung cancer has been known for many years, the precise mechanism by which smoking gives rise to cancer has not.

However, in 1996, scientists from Texas and California found at least part of the—pardon the expression—"smoking gun" of smoking-promoted cancer. They showed that a chemical in tobacco smoke damages the *p53* tumor suppressor gene in lung cells, and that damage is found in many lung tumors.

The process by which cancer develops is enormously complex and still not well understood. New discoveries are being made almost every day, however, and the prospects for new and better treatments down the road—many of them based on our increasing knowledge of the genetics behind this dreaded disease—are increasingly bright.

Quiz

1. What did Francis Peyton Rous determine had caused the tumors in the chickens he was studying?
   (a) a virus
   (b) a bacteria
   (c) crowded coop conditions
   (d) the farmer's smoking habit

2. The immortalization of cancer cells means
   (a) cancer cells can never be killed.
   (b) cancer cells become independent organisms.
   (c) cancer cells can divide indefinitely.
   (d) cancer cells can infect other organisms.

3. Metastasis is
   (a) a method of freezing cancer cells for study.
   (b) cancer cells' ability to move independently and penetrate other tissues.
   (c) a gene that makes cancer cells effectively immortal.
   (d) a sequence of "junk DNA" specific to cancer cells.

4. A population of cells growing out of control is called a
   (a) neopsis.
   (b) neologism.
   (c) neocrology.
   (d) neoplasm.

5. Genes that promote cancer are called
   (a) narcogenes.
   (b) wild genes.
   (c) oncogenes.
   (d) oncosomes.

6. The non-cancer-causing versions of the genes mentioned in question five are called
   (a) prenarcogenes.
   (b) tame genes.
   (c) protooncogenes.
   (d) noncosomes.

7. The first tumor suppressor gene was isolated from
   (a) retinoblastoma.
   (b) colon cancer.
   (c) bladder cancer.
   (d) Hodgkin's lymphoma.

8. The tumor suppressor gene that shows up in approximately half of all U.S. cancers is called
   (a) *u2*.
   (b) *p51*.
   (c) *6pac*.
   (d) *p53*.

9. Mutagens that turn cells cancerous are called
   (a) tumorgens.
   (b) carcinogens.
   (c) oncogens.
   (d) promogens.

10.. Some cancer-causing mutagens have to be activated by an enzyme inside the organism. This process is called
   (a) enzymatic activation.
   (b) interorganismal activation.
   (c) metabolic activation.
   (d) seroactivation.

Bacteria—A Different Way of Doing Things

Until now, our focus has been on eukaryotic genetics, mainly because everyone reading this book is (I'm pretty sure) a eukaryote. But now let's devote some time to the prokaryotes, not least because, as you'll see when we start talking about genetic engineering, they're central to the genetics revolution that is remaking the world.

Plus, of course, there are more of them than there are of us.

Characteristics of Bacteria

The eukaryotes we've focused on until now may be either singlecelled or multicelled, but however many cells they have, those cells have their genetic material isolated from the rest of the cell by a nuclear membrane.

Prokaryotes, on the other hand, are almost always singlecelled, and within those cells, genetic material is not confined to a nucleus. (Nor do they often have other membrane-bound organelles, as eukaryotic cells are wont to do.)

All bacteria are prokaryotes, and they can be further divided into two groups:

1. *Eubacteria.* These "true bacteria" include most of the bacteria we're most familiar with, such as *E. coli.*
2. *Archaea.* Thought to predate the *Eubacteria,* evolutionarily, the *Archaea* include exotic organisms such as the *methanogens* (bacteria that produce methane).

Bacteria are also often classified by shape: *bacilli* (rod-shaped), *cocci* (spherical), *spirilla* (spiral), *spirochetes* (helical), and *branched.* However, because they're so small, they're more often studied as colonies or populations than as individual cells.

Another way bacteria can be classified is as either *Gram-positive* or *Gram-negative.* The plasma membrane of most bacteria is surrounded by a wall that contains a chemical called *peptidoglycan.* In Gram-positive bacteria, the wall includes a thick layer of peptidoglycan. In Gram-negative bacteria, the peptidoglycan layer is thinner, but it's compensated for by an extra outer membrane. (Penicillin kills bacteria by disrupting their synthesis of peptidoglycan.)

As you might guess from their lack of a proper nucleus, the process of cell division is significantly different in bacteria than it is in eukaryotes like you and me.

To begin with, the bacterial chromosome is a single, ring-shaped, double-stranded DNA molecule, which only sometimes is combined with proteins to form a chemical similar to eukaryotic chromatin. Unlike our chromosomes, it doesn't condense during cell division. It also has no centromere.

After the bacterial chromosome replicates, the two copies move apart as the bacteria elongates. A new cell wall is synthesized between the copies, and *voila!,* two new bacteria. This method of reproduction is called *binary fission,* and bacteria are much faster at it than eukaryotic cells: Under ideal conditions, bacteria can divide once every 20 minutes, whereas it may take a full day, or even two, for eukaryotic cells to divide.

Let's look at the replication process in bacteria in more detail.

DNA Replication and Cell Division

As you'll recall, eukaryotic chromosomes usually have many ori (origin) sites at which DNA replication may begin. Bacterial chromosomes, in contrast, have only one. DNA replication usually proceeds in both directions at once around the ring-shaped chromosome, creating two replication forks.

The unwinding of the DNA helix in both directions would cause the chromosome as a whole to twist in the same direction as the helix and become so tightly wound that replication would cease—if not for the action of an enzyme called *DNA gyrase.* One of a group of enzymes called *topoisomerases* that can change the shape of DNA molecules, DNA gyrase prevents this *positive supercoiling* by twisting DNA in the opposite direction.

It's thought that DNA gyrase does that by cutting the DNA, holding onto the cut ends so the molecule doesn't rotate, passing an intact length of the molecule through the opening, and then sealing up the break again (see Fig. 9-1).

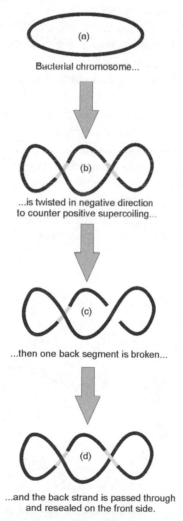

Fig. 9-1. The bacterial enzyme DNA gyrase can change the twist of the chromosome of a bacteria by passing one strand through another. Above, the circular bacterial chromosome (a) is first twisted (b), then a cut is made through which an intact part of the strand can be passed (c), then the break is resealed on the other side (d).

Localized regions of DNA on the chromosome will also briefly unwind, forming single-stranded *bubbles,* to relieve the strain of all this twisting, then reform, in a repeating process rather like breathing. Bubbles get their energy from the increased molecular vibration caused by heat; as the temperature increases, so does the number of bubbles.

However, at the replication forks themselves, the DNA is unwound not by heat but by an enzyme called *helicase.* Once the single strands have been unwound, single-stranded DNA-binding proteins (*SSB proteins*) keep the unwound DNA strands from rebinding with each other.

The enzyme *primase* uses a region on each unwound strand as a template for short RNA strands called primers, which the enzyme *DNA polymerase* requires in order to begin duplicating the DNA.

Of all the bacteria, the species most studied is undoubtedly *E. coli.* Scientists have identified three DNA polymerases in *E. coli,* called (rather unimaginatively but simply) *pol I, pol II* and *pol III.* Pol III is the one that takes charge of the main duplication effort. Pol I fills in any gaps left by Pol III. DNA ligase seals up any nicks. The function of Pol II still isn't completely understood.

About halfway through replication, the chromosome looks a bit like the Greek letter *theta,* so this process is called *theta replication* (see Fig. 9-2).

However, there's another method DNA chromosomes can use to replicate all or part of themselves. It produces a shape rather like the Greek letter *sigma,* and so is called (you guessed it!) *sigma replication.*

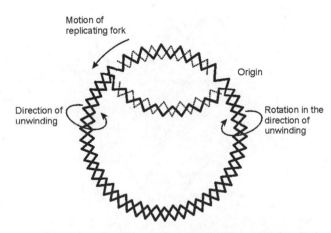

Fig. 9-2. About halfway through replication, the bacterial chromosome looks something like the Greek letter *theta.* The dotted lines indicate newly synthesized DNA. DNA gyrase, as shown in Fig. 9-1, keeps the unreplicated parts from overwinding despite the tension placed on them by the unwinding of the part being replicated.

Sigma replication is used when one bacterium needs to transfer a straight piece of DNA to another bacterium through conjugation (more on that later). It may also take place at the behest of a bacteria-infecting virus, or phage, which has taken over the bacteria's genetic machinery to reproduce copies of itself and needs straight segments of DNA to do so.

In sigma replication, a nick occurs in one strand of a DNA double helix, and helicase and SSB proteins establish a replication fork at that point. As the leading strand is replicated, the template for the lagging strand is displaced, and replicated in discontinuous *Okazaki fragments* (as described in Chapter 3). This is essentially the same way linear DNA is replicated in eukaryotes. The result is a circle with a linear tail (see Fig. 9-3).

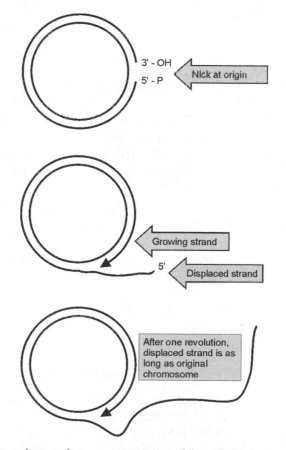

Fig. 9-3.   When a bacterium needs to create a segment of linear DNA, it uses sigma replication, sometimes called rolling circle replication (for obvious reasons).

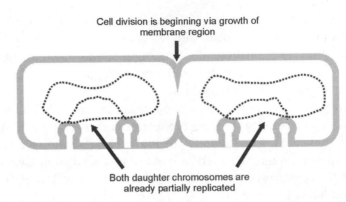

Cell division is beginning via growth of
membrane region

Both daughter chromosomes are
already partially replicated

Fig. 9-4. Bacteria can replicate quickly because, even before cell division is complete, the
chromosomes of the daughter cells may already be in the process of replicating
to create the next generation.

If the entire genome is being copied, the circle may revolve several times,
creating linear copies of itself which have "sticky ends" (single-stranded complementary ends) that can then be joined by DNA ligase, creating new circular
chromosomes.

As a bacterial chromosome replicates, it attaches to an infolded hollow space,
or *invagination,* in the cell membrane at each replication fork. Once replication
is complete, the bacteria elongates by adding material to the area between the
two attachment points—so, of course, the duplicated chromosomes draw apart
from each other. Eventually, new cell membranes are synthesized between the
two parts of the original cell and fission is complete.

One reason bacteria can replicate so fast under ideal conditions is that cells
may sometimes contain two to four chromosomes in various stages of replication (see Fig. 9-4).

Bacterial Transcription and Translation

All RNA molecules involved in transcription in bacteria (messenger RNA,
ribosomal RNA, and transfer RNA) are synthesized by the same enzyme,
RNA polymerase. Each gene has a promoter region preceding it, which marks
where transcription is to begin.

Bacterial RNA polymerase consists of five different polypeptide chains, one of which, called the *sigma subunit,* is involved in initiating transcription. It recognizes (and binds to) a sequence of single-strand DNA in the promoter region 10 bases upstream of the first base to be copied (a region called the *Pribnow box* or the *–10 box*). At 35 bases upstream (labeled simply *–35,* the first base to be copied is labeled the +1 base and any sequence upstream of it is given a negative number.

Sigma's binding to this region aligns the remainder of the enzyme (called the *core enzyme*) correctly, so that transcription begins where it's supposed to begin. Once transcription begins, sigma dissociates from the core enzyme. It's then free to help initiate transcription again.

Transcription stops when RNA polymerase runs into a DNA sequence called a *terminator.* The RNA molecule then dissociates itself from the DNA.

A single RNA transcript in bacteria can contain the coding for more than one gene, in which case it's called *polycistronic* or *polygenic.* (This is rare in eukaryotes because, unlike prokaryotes, their genes are usually separated by long introns—stretches of noncoding DNA sequences.) The resulting RNA molecule will have a leader sequence at the 5′ end, a trailer sequence at the 3′ end, and spacing sequences (that serve no other function) separating the gene sequences. Often, the genes transcribed in this way have related functions and were transcribed under the control of the same promoter, in which case they're said to make up an *operon.*

Unlike eukaryotes, prokaryotes do not have a nuclear membrane separating their DNA from the rest of the cell. That means ribosomes are free to begin translation as soon as transcription is complete.

Translation in bacteria has three major steps:

1. *Initiation.* In the initiation phase, rRNA binds to a special sequence of mRNA that locks it into the correct place to begin translation. As in eukaryotes, the initiation codon near the 5′ end of an mRNA molecule is AUG, which codes for the amino acid methionine. A *formyl group* (chemical formula CHO) attaches to the molecule of methionine once the amino acid is attached to its tRNA. An enzyme called *deformylase* removes that formyl group from some polypeptide chains shortly after their synthesis begins. Another enzyme, called *aminopeptidase,* may also remove the methionine from the end of the chain, so that not all bacterial proteins have either formyl methionine or methionine at their starting end.
2. *Elongation.* In this phase, the ribosome begins assembling a polypeptide chain, one amino acid at a time, just as it does in eukaryotes. Several ribosomes can translate the same mRNA simultaneously. A collection of multiple ribosomes like this is called a *polyribosome.*

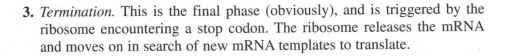

3. *Termination.* This is the final phase (obviously), and is triggered by the ribosome encountering a stop codon. The ribosome releases the mRNA and moves on in search of new mRNA templates to translate.

Some bacteria create proteins that are intended to be released through the cell's membrane (the toxins produced by many of the bacteria that cause food poisoning, for instance). These usually contain 15 to 30 additional amino acids at one end that anchor themselves in the membrane, holding the protein in place while it is synthesized and extruded through the membrane. Once the protein is complete, an enzyme called a *signal peptidase* clips the signal sequence off from the rest of the protein, releasing the protein outside the cell.

Genetic Recombination of Bacteria

The exchange of DNA between independent bacterial cells is called *horizontal gene transfer.* There are three main mechanisms for it: *transformation, conjugation,* and *transduction.*

CROSS-REFERENCE

For more on genetic recombination, see Chapter 13.

TRANSFORMATION

Transformation is the transfer of naked DNA (which usually originated in another cell) into a bacterial cell. Naked DNA is released into the environment when a bacterial cell undergoes *cellular lysis*—that is, it ruptures.

Only some bacterial cells can incorporate naked DNA. Those that can are said to have *competence.* Not all species of bacteria are competent, and even those that are are usually competent only during a limited portion of their life cycle. During that time, they produce proteins called *competence factors* that modify the cell wall so it can bind to foreign DNA fragments and help the cell absorb and incorporate that DNA.

Foreign DNA fragments that pass into a new bacteria by any means (transformation, conjugation, or transduction) are called *exogenote.* The native DNA of the receiving cell is called the *endogenote.* A genetic exchange that transfers part of the genetic material from one bacteria to another is called *meromixis*. A cell containing unincorporated naked DNA is called a *merozygote*.

As a double-stranded fragment of naked DNA passes through the bacteria cell well, one of the strands is degraded. This generally makes the exogenote unstable, to the point that it will soon disintegrate unless it is incorporated into the endogenote.

It's thought that the exogenote is coated with a protein that helps it find a complementary region on the endogenote and insert itself there (replacing part of the endogenote in the process, of course).

If the exogenote contains a gene that differs from the matching gene in the strand of the endogenote it has replaced, there will obviously be some mismatched base pairs. The cell clips out the mismatched base from the endogenote and uses the exogenote as a template to replace it. In this way, new genetic information is incorporated into the bacteria. (For this process to work at all, of course, the donor and recipient cells must usually belong to either the same species or two closely related ones.)

If the exogenote contains two or more closely linked genes, both of them may be transformed into the recipient cell. In that case, the cell is said to be *cotransformed*.

CONJUGATION

Conjugation is the closest bacteria come to having sex. Two cells of opposite mating type form a cytoplasmic bridge through which some genetic material is passed from one cell to the other.

That genetic material typically takes the form of either an *episome* or a *plasmid*.

An episome is a genetic element that may either be a small circular DNA molecule that is separate and apart from the bacterial chromosome and replicates autonomously, or a distinct sequence integrated into the chromosome. A plasmid is also a small, circular, autonomously replicating circle of DNA, typically between 1/10th and 1/100th the size of the whole chromosome. Unlike episomes, plasmids are never incorporated into the chromosome. (Note, however, that the word plasmid is now often used as an umbrella term to cover episomes and what were originally defined as plasmids.)

Conjugation has been studied most closely in *E. coli* (naturally). In some strains of *E. coli,* some individuals carry the *F plasmid,* which gives them the ability to manufacture a protein called *pilin*. This protein is used to build a conjugation tube, or *pilus,* between these F+ (or male) individuals, and those without the F plasmid (F–, or females). Once the tube is built, one strand of the F plasmid breaks and it is copied into the female cell by means of sigma replica-

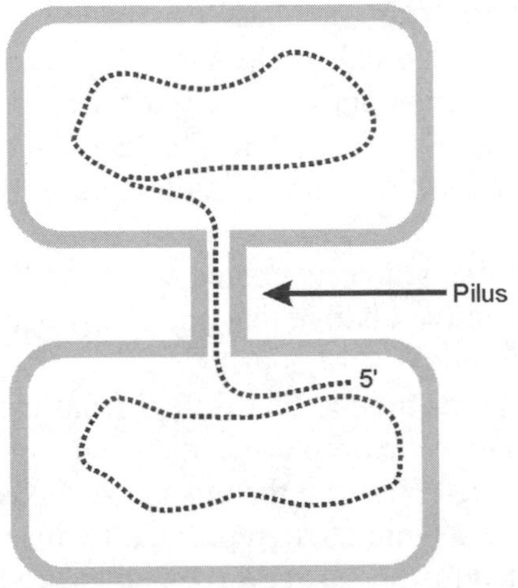

Fig. 9-5. In *E. coli* conjugation, a single strand of DNA is transferred from the F+, or male, cell into the F-, or female, cell where it replicates and changes the cell's "gender."

tion (see Fig. 9-5). As a result, the female cell gains an F plasmid and becomes F+, or male.

Most of the time, the F factor remains extra-chromosomal. In about 1 in 10^5 cells, however, it manages to integrate itself into the recipient's chromosome. That cell is then called an *Hfr* (high frequency of recombination) cell. Having the F factor in its chromosome makes it capable of transferring large portions of its genome to the other cell. Since the conjugation usually ruptures before the complete genome can be transferred, the recipient cell usually does not receive the F factor and thus remains F– rather than becoming an Hfr cell itself.

Under normal conditions, bacteria without any plasmids can survive just fine. When bacteria are put under environmental stress, though, plasmids become important, because some plasmids carry genetic information that give their host cells certain advantages. Plasmids carry genes that give cells antibiotic resistance, for instance, or resistance to other toxins. Plasmids that confer some kind of resistance are called *R plasmids* or *R factors*.

TRANSDUCTION

Transduction is the transfer of material from one bacterium to another via a bacterial virus, or *phage*. Except for the vector, it doesn't vary much from the mechanisms discussed above.

CROSS-REFERENCE

Viruses are discussed in detail in Chapter 11.

Regulation of Gene Activity in Bacteria

In prokaryotes, as in eukaryotes, not all genes are active at the same time. Even when they are active, their output sometimes needs to be controlled so that only the correct amount of a protein is produced.

Proteins that are made all the time are said to be synthesized *constitutively*. Their genes are only regulated by the affinity of the genes' promoters for RNA polymerase—the more affinity those promoters have for the product, the more constantly their gene codes for it will be produced.

Other genes are under the control of regulatory proteins. These aren't enzymes, usually, but interact with promoters to regulate transcription. Just as in eukaryotes, there are *repressors* and *activators*. Repressors bind to a site called the *operator* within an *operon*. This prevents transcription of all the genes in the operon. Since the absence of the repressor is necessary for an operon to be active, this is called *negative control*.

Proteins that must be present in order for an operon to activate are called (as you'd expect) *activators*. They may bind to sites in the operon's promoter or sequences far from the operon called *enhancer sites*. Since the presence of the activator is necessary for the operon to activate, this is called *positive control*.

All kinds of stimuli can activate these control mechanisms, ranging from tiny molecules like sugars and amino acids to large substances like hormone molecules. Substances that turn on gene transcription are called *inducers;* those that turn transcription off are called *corepressors*.

Genes that can be activated by an inducer are called (go figure) *inducible genes,* and they're usually involved in reactions that break down other substances. Genes that can be shut off by corepressors are called *repressible genes,* and they're usually involved in processes that create substances, such as amino acids.

So, the possible variations are negative, inducible control; negative, repressible control; and positive, inducible control (no positive, repressible control is known).

An Example of Gene Regulation in *E. coli*

The classic example of gene regulation in bacteria is drawn from, yet again, *E. coli*.

E. coli contains an enzyme called *β (beta)-galactosidase,* which breaks down lactose, the sugar in milk, into two simpler sugars: galactose and (*E. coli*'s preferred nutrient), glucose.

E. coli only synthesizes *β*-galactosidase if lactose is the only sugar available. Several DNA sequences that precede the 5' end of the sequence that codes for *β*-galactosidase regulate the transcription of the enzyme (see Fig. 9-6). RNA polymerase binds to one of these sequences, the promoter. A sequence called the operator lies between the promoter and the start of the *β*-galactosidase sequence. The operator interacts with a repressor sequence. (The shape of the repressor and the length and organization of the operator make it unlikely the repressor will bind anywhere else but within the operator.) Normally, binding of the repressor to the operator prevents RNA polymerase from beginning transcription.

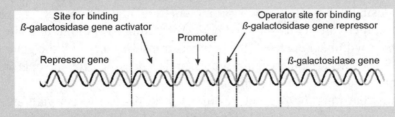

Fig. 9-6. These DNA segments regulate transcription of the *β*-galactosidase gene in *E. coli*.

If, however, *E. coli* is supplied with lactose, the sugar binds to the repressor protein. This alters its shape. That, in turn, prevents it from binding to the DNA. And that, in turn, enables RNA polymerase to transcribe *β*-galactosidase—and the lactose to be used as an energy source by the *E. coli*.

That's an example of negative control.

This same mechanism also provides an example of positive control. The synthesis of *β*-galactosidase also requires the presence of a specific activator protein. This protein functions only when joined by a special small molecule that appears in the cell only when no glucose is available to the cell—when it's starving, in other words. Once that small molecule appears, it forms a complex with the activator, which then binds to a short segment of DNA near the promoter–operator region and enhances the ability of RNA polymerase to transcribe the *β*-galactosidase gene. Thus, the full expression of *β*-galactosidase requires both the presence of lactose and the absence of glucose.

Although the repressor/activator mechanism is the most common form of gene regulation in bacteria, it's not the only one. In some cases, the protein produced by a gene regulates that gene's transcription itself, through a feedback regulatory system—if too much of the protein is present, it inhibits production of more until the level falls again.

There are, in fact, regulatory mechanisms for just about every step of the protein synthesis process in prokaryotes just as there are in eukaryotes (as was discussed back in Chapter 5).

Prokaryotes may do some things very differently—but they also do some things the same.

One of the biggest differences between eukaryotes and prokaryotes—and its importance to understanding genetics—is discussed in the next chapter.

Quiz

1. The main difference between prokaryotes and eukaryotes is
   (a) eukaryote cells have a nucleus; prokaryote cells don't.
   (b) eukaryotes practice photosynthesis; prokaryotes don't.
   (c) prokaryotes live longer.
   (d) prokaryotes are much rarer than eukaryotes.

2. The bacterial chromosome is
   (a) a single linear strand of DNA.
   (b) two double-helix strands twisted together.
   (c) a single ring-shaped DNA molecule.
   (d) scattered throughout the bacteria in fragments.

3. The function of DNA gyrase is to
   (a) twist bacterial DNA very tightly together.
   (b) keep bacterial DNA from twisting so much that it can't replicate.
   (c) make bacteria spin in place to fight off viruses.
   (d) attract other bacteria for conjugation.

4. Under ideal conditions, bacteria can replicate
   (a) once a day.
   (b) twice a day.
   (c) every two hours.
   (d) every 20 minutes.

5. A polycistronic length of mRNA contains codes for
   (a) sugar.
   (b) more than one gene.
   (c) cell walls in Gram-positive species.
   (d) disease toxins.

6. Bacteria that can incorporate naked DNA are called:
   (a) hungry.
   (b) brave.
   (c) competent.
   (d) modest.

7. When two bacteria join together so that one can give the other a copy of some of its genetic information, the process is called
   (a) confrontation.
   (b) conjugation.
   (c) confabulation.
   (d) commitment.

8. A small DNA molecule that is outside of the bacterial chromosome and replicates autonomously is called a
   (a) plasmid.
   (b) plasmon.
   (c) extrasome.
   (d) parasite chromosome.

9. Proteins that must be present in order for genes to activate are called
   (a) activists.
   (b) accountants.
   (c) provokers.
   (d) activators.

10. Substances that trigger the shutdown of gene transcription are called:
    (a) co-conspirators.
    (b) co-repressors.
    (c) conquerors.
    (d) controllers.

10

Organelles—Genetics Outside the Nucleus

On New Year's Day 1987, a paper appeared in the magazine *Nature* that did something quite rare in the annals of research: it sprouted a whole new branch of science.

"Mitochondrial DNA and human evolution," by Allan Wilson, Rebecca Cann, and Mark Stoneking, reported research that indicated that almost everyone alive today, no matter where they live in the world, can be genetically linked to one woman who lived about 200,000 years ago, probably in Africa.

The paper effectively launched the field of evolutionary genetic research. What makes it worth mentioning off the top of this chapter is that the DNA that Wilson, Cann, and Stoneking studied to reach their remarkable conclusion wasn't the DNA we've mostly talked about so far—located in the 23 chromosomes of the human cell's nucleus. Instead, they studied DNA that's found outside the cell nucleus, in the tiny organelles called *mitochondria*.

CROSS-REFERENCE

The genetics of evolution is discussed in more detail in Chapter 14.

As you'll recall from Chapter 2, organelles are small, membrane-enclosed structures that perform a number of vital functions inside cells. Two of them (not counting the nucleus, which is also an organelle and is, of course, crammed with DNA) contain their own genetic material: mitochondria and chloroplasts.

In this chapter we'll take a look at these two other DNA-containing parts of the cell and their importance for cells and evolutionary genetic research.

Mitochondria

Mitochondria are the second-largest organelles in the cell (after the nucleus). They contain enzymes necessary for the production of adenosine triphosphate (ATP), the basic energy source for biochemical reactions within cells (see Fig. 10-1). Essentially, they're the cell's—and therefore the whole organism's—power-generating plants.

Mitochondria share a lot of characteristics with the prokaryotes we looked at in the last chapter. Like prokaryotes, mitochondria generally have a circular, double-stranded DNA genome. (However, there are some protozoa, including

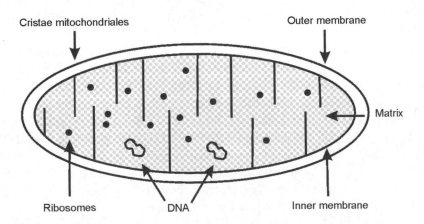

Fig. 10-1. This diagram illustrates the structure of a mitochondrion. It has two membranes, separated by a space. Inside the inner membrane is the dense, somewhat granular *matrix,* which also contains the mitochondrial DNA and ribosomes. The *cristae* are shelves (in a few species, tubes) projecting inward from the inner membrane that give the mitochondria the compartmentalization it needs to carry out its biochemical tasks.

the well-known *Paramecium*, that have linear mitochondrial DNA.) Also, like that of prokaryotes, the DNA in mitochondria is not tucked away inside a nuclear membrane. Protein synthesis in mitochondria follows the same general outline as in prokaryotes, too. Not only that, mitochondria grow and then split in two, in a form of binary fission.

These similarities to bacteria may have a very simple explanation: Mitochondria may actually have been bacteria, eons ago in evolutionary time.

The prevailing theory for the origin of all organelles, in fact (not just those that still have DNA of their own), is that they arose from symbiotic relationships between early bacterial cells. This is called the *endosymbiont* theory.

According to this theory, mitochondria came about because a primitive nucleated cell of a type that couldn't use oxygen to generate energy (termed the *urkaryote*) engulfed a bacteria that could (called the *progenote*). The progenote managed not only to survive but to replicate inside the cytoplasm of the urkaryote, so that copies of itself remained inside subsequent copies of the urkaryote. This accidental relationship eventually evolved into a true symbiosis, where neither type of bacteria could survive on its own.

As the relationship evolved, the progenote gave up many of its genes to the nucleus of the urkaryote—which is why today's mitochondria are not self-sufficient. Although their genome codes for most of the components of their own protein-synthesizing system (rRNA, tRNA, and so on), many of the enzymes and other proteins that the mitochondria needs to carry out its functions for the cell are synthesized by the nuclear genome, then transported into the organelle.

Mitochondrial ribosomes are usually smaller than those of the cell proper. This makes them sensitive to some antibiotics and other substances that don't affect the cell's cytoplasmic ribosomes.

The length of the mitochondrial genome varies substantially from species to species. In fungi such as yeast, it's often quite large, as much as 86,000 base pairs (although much of that DNA is thought to be noncoding). In multicellular animals it averages just 16,000 base pairs (in humans, 16,569, to be precise). Within animals, all mitochondria typically code for the same 37 proteins. There are two rRNA genes, 22 tRNA genes, and 13 genes that code for proteins involved in respiration, DNA replication, transcription, and translation.

Each mitochondrion has several nucleoid regions in which the same DNA can be found, which means any given cell has a great many copies of mitochondrial DNA: If each mitochondria contains just five copies of its DNA, and the cell contains 200 mitochondria, after all, that's 1000 copies of the mitochondrial genome right there. Multiple copies may be important because mitochondria cannot repair damaged DNA, which means the mutation rate in mitochondrial DNA is much higher than in nuclear DNA.

Table 10-1. Changes to the Genetic Code Found in Mitochondria

Organism	Codon	Meaning in Mitochondrial DNA	Usual Meaning in Nuclear DNA
Many	UGA	Tryptophan	Termination
Mammal	AGA, AGG	Termination	Arginine
Mammal	AUA	Initiation	Isoleucine
Fruit fly	AUA	Initiation	Isoleucine
Yeast	AUA	Elongation	Isoleucine
Yeast	CUA	Threonine	Leucine
Fruit fly	AGA	Serine	Arginine

Although transcription and translation of mitochondrial DNA work pretty much the same as in bacteria, there are some differences in the genetic code, some of which vary from species to species (see Table 10-1). These differences have probably arisen because of the higher rate of mutation in mitochondrial DNA.

Chloroplasts

Chloroplasts are also thought to have originated as separate organisms, but much later than mitochondria. The generally accepted theory is that chloroplasts were originally photosynthetic *cyanobacteria* (aka. blue-green "algae") engulfed by fully nucleated, mitochondria-bearing eukaryotic cells. As had the mitochondria before them, these cyanobacteria developed a symbiotic relationship with the engulfing cells, and eventually turned into what we see today: the chlorophyll-bearing photosynthesis-based powerhouses of the plant kingdom (see Fig. 10-2).

Most plant cells contain many chloroplasts, although there are species (unicellular algae, for instance) that contain only one chloroplast per cell. A typical number of chloroplasts in a cell might be 40 to 50, each of which may contain 40 to 80 copies of the DNA molecule, which cluster together and are thought to attach to the inner membrane.

Chloroplast DNA is larger than mitochondrial DNA. It ranges from 120,000 to 150,000 base pairs in length and contains anywhere from 46 to 90 genes. Most

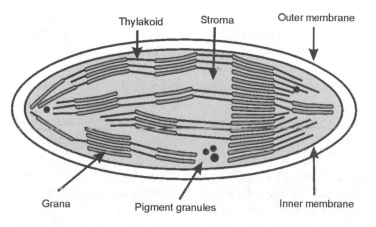

Fig. 10-2. This diagram illustrates the structure of a chloroplast. Like a mitochondrion, it has two membranes, separated by a space. The inner membrane contains special proteins called transporters that regulate the passage in and out of small molecules like sugars and the proteins synthesized by the cell's nuclear DNA for use within the chloroplast. The inside is filled with a system of *thylakoid* membranes that contain the proteins, including chlorophyll, that carry out the light reactions of photosynthesis, and the fluid *stroma* which contains the substances necessary for the dark reactions of photosynthesis—and several copies of the chloroplast's DNA.

of the proteins it codes for are involved in photosynthesis. The others relate mainly to the chloroplast's own functions of replication, division, transcription, and translation. Like mitochondria, chloroplasts require some proteins coded by nuclear DNA in order to function.

Plants from liverworts on up the evolutionary ladder all have essentially the same chloroplast genome.

Organelle Inheritance

Both mitochondria and chloroplast DNA have one characteristic that makes them quite different from nuclear DNA: They're inherited almost exclusively from the female parent.

The male gamete, or at least that part of it that is involved in fertilization, simply doesn't have very many organelles. In animals, mitochondrial inheritance is almost 100 percent maternal; in plants, it's estimated that in two-thirds of species the inheritance of chloroplasts is strictly maternal.

It is possible for the DNA in organelles to influence traits in the organism as a whole. One example is found in yeast. Yeast cells, called *petites,* are slow-growing because of the lack of normal activity from a respiratory enzyme associated with mitochondria. When petites are mated with wild-type yeast, the result is a fertile wild-type diploid cell which, under proper conditions, will reproduce sexually. The resulting four offspring are all wild-type. The petite trait won't appear even with extensive back-crossings, as would be expected under normal Mendelian rules. This indicates that the petite trait arises from a deficiency in mitochondrial DNA rather than nuclear DNA.

Problems with the mitochondria can also cause illness in humans. If mitochondria aren't functioning properly, they provide less and less energy—which results in cell injury and death. If the problem is widespread in the body, whole systems can fail, producing painful, debilitating symptoms that can be life-threatening.

The Past Through Mitochondria

The fact that mitochondrial DNA inheritance is strictly maternal in humans is what has made it attractive as a means of studying human evolution.

In the paper mentioned at the start of this chapter, Wilson, Cann, and Stoneking reported on a thorough comparison of mitochondrial DNA taken from 147 people from five different geographic populations: African, Asian, Australian, Caucasian, and New Guinean. They were looking for variations in the mitochondrial DNA in about nine percent of the total mitochondrial genome.

On the theory that two sequences that were only slightly different had to have diverged more recently than two sequences that were very different from each other, the researchers grouped the sequences according to the degree of variation among them. They ended up with two large groups: the sequences from the African area, and all the rest.

That indicated that the oldest sequences were those from Africa, while the non-African sequences all had more recent and diverse origins.

Using the estimate that mitochondrial mutations accrue at a rate of two to four percent per million years, Wilson *et al* concluded that the mitochondrial DNA they had examined had arisen from a single woman who lived about 200,000 years ago, probably in Africa.

More recent research by Douglas Wallace of Emery University in Atlanta has suggested that this "African Eve" had 18 descendants—each with a distinct mitochondrial genome—that spread to different regions of the globe.

Since then, mitochondrial DNA has given scientists other insights into human history. For example, in 1997 a team of researchers led by Svante Pääbo, a molecular anthropologist, reported that they had managed to extract enough intact genetic material from an original Neanderthal skeleton—379 bases of mitochondrial DNA, to be precise—to conclude that *Homo sapiens* are not descendants of Neanderthals, but merely cousins.

Pääbo came to that conclusion because, whereas the stretch of mitochondrial DNA he examined typically varies at eight locations among individual humans, he found 27 differences between the Neanderthal DNA and human DNA. This suggests Neanderthals and humans diverged around half a million years ago. (Later research by William Godwin of the University of Glasgow verified Pääbo's findings.)

Mitochondrial DNA has also proved valuable in research aimed at periods a little more recent. Bryan Sykes, a geneticist at Oxford University's Institute for Molecular Medicine, has been using mitochondrial DNA to study human variation and migration. He has traced the ancestry of modern Europeans to seven ancestral women. (He's even set up a company that allows individuals to submit DNA samples and discover which of these "Seven Daughters of Eve" they're most closely related to.)

Another Tool in the Evolutionary Geneticist's Toolbox

Mitochondrial DNA isn't the only form of DNA that can be used by geneticists studying human evolution and migration. Just as the study of mitochondrial DNA sheds light on the inheritance of genes through the maternal line, the Y chromosome sheds light on the inheritance of genes through the paternal line—since, by definition, it's the presence of a Y chromosome that makes men male.

Research by David Page and Bruce Lahn of the University of Chicago indicates that the mammalian Y chromosome is actually the result of a mutation about 300 million years ago. (In our reptilian ancestors, sex was determined by environmental conditions such as temperature, not by genes.)

Millions of years of subsequent mutations have rearranged and deleted genes from the Y chromosome to the point where the chromosome is hardly able to share any genes during meiosis with the X chromosome, even though they're partners in the 23rd pair of human chromosomes.

That means that the only significant source of change in the Y chromosome is mutation. By comparing specific segments of Y-chromosome DNA that are most prone to variation, taken from men from various regions, a historical picture that complements the mitochondrial one can be drawn.

The genomes of both mitochondria and chloroplasts are much smaller than the genomes found in either bacteria or the nuclei of eukaryotic cells. But they're certainly not the smallest genomes known.

Those belong to viruses—which we'll examine in the next chapter.

Quiz

1. The cellular energy source provided by mitochondria is abbreviated
   (a) STP.
   (b) CBC.
   (c) ATP.
   (d) PDQ.

2. The theory that organelles have evolved from bacteria engulfed by a long-ago cell is called
   (a) the exobacterial theory.
   (b) the amoebic theory.
   (c) the Big Gulp theory.
   (d) the endosymbiont theory.

3. Human mitochondrial DNA has approximately
   (a) 1600 base pairs.
   (b) 16,000 base pairs.
   (c) 160,000 base pairs.
   (d) 1,600,000 base pairs.

4. The mutation rate in mitochondrial DNA is
   (a) lower than in nuclear DNA.
   (b) higher than in nuclear DNA.
   (c) about the same as in nuclear DNA.
   (d) non-existent.

5. Chloroplasts probably originated as
   (a) cyanobacteria.
   (b) viruses.
   (c) *E. coli* bacteria.
   (d) yeast cells.

6. Compared to mitochondrial DNA, chloroplast DNA is typically
   (a) shorter.
   (b) longer.
   (c) about the same.
   (d) more twisted.

7. Both mitochondrial DNA and chloroplast DNA are overwhelmingly inherited
   (a) from the maternal side.
   (b) from the paternal side.
   (c) from both parents.
   (d) from symbiotic bacteria.

8. Mitochondrial research suggests we're all descended from a woman who lived
   (a) 10,000 years ago in Scotland.
   (b) 100,000 years ago in Russia.
   (c) 200,000 years ago in Africa.
   (d) 500,000 years ago in China.

9. Mitochondrial evidence suggests
   (a) Neanderthals are our direct ancestors.
   (b) Neanderthals are still alive.
   (c) Neanderthals never existed.
   (d) Neanderthals and humans last shared a common ancestor 500,000 years ago.

10. To study patterns of human migration through the paternal line of inheritance, researchers turn to
   (a) the Y chromosome.
   (b) the X chromosome.
   (c) *The X Files*.
   (d) Robert E. Howard's *Conan* stories.

Viruses—
Hijacking Heredity

In 1886 German scientist Adolf Mayer spent a great deal of time and effort trying to isolate the bacteria that caused a disease of the tobacco plant called tobacco mosaic disease. According to the still-brand-new germ theory of disease, some sort of microorganism had to be responsible.

He failed to find it.

Six years later, Russian scientist Dmitri Iwanowski conducted his own experiments, and decided that whatever was causing the tobacco mosaic disease couldn't be a bacteria. He thought this because the infectious agent couldn't be filtered out by even his finest porcelain filters, whose pores were much too small for any known bacteria to pass through. He thought the infectious agent must be a toxin of some kind produced by some kind of unculturable bacteria—either that, or his filters had leaks in them.

Although apparently unaware of Iwanowski's work, Dutch scientist Martinus Beijerinck replicated Iwanowski's filtration experiments in 1898. His conclusions were different, however. Because an infinite number of healthy plants could be infected by the sap of a diseased plant—even after it had been filtered—he concluded that whatever was in the fluid that passed through the filters must be an infectious, living agent. Other experiments confirmed that it could not be a microbe. He needed a new word for whatever it was, and so he called it a *virus*.

Almost at once, other viruses were discovered to be responsible for other diseases. The virus responsible for foot-and-mouth disease in cattle was isolated in 1898, and the virus responsible for yellow fever was discovered in 1900.

Fifteen years later, viruses that infect bacteria, called *bacteriophages* ("eaters of bacteria") were discovered. But virus particles are so small that no one actually saw one until the first electron microscopes—with magnification capabilities 1000 to 2000 times greater than optical microscopes—were developed in the 1930s.

Today, the International Committee on the Taxonomy of Viruses recognizes about 3000 species of viruses—and those are thought to be just a fraction of what exists. The ones we tend to be most aware of are the ones that cause human diseases, ranging from the common cold to chicken pox to influenza to AIDS. In fact, any time we get the sniffles or an upset stomach, we're likely to say, "I must have caught that virus that's going around."

So what is a virus? And how does it fit into the grand scheme of genetics?

What Is a Virus?

A virus is an ultramicroscopic, obligate, intracellular parasite incapable of autonomous replication. *Ultramicroscopic* means, as you might expect, very small. Viruses are much, much smaller than either prokaryotic or eukaryotic cells. A typical virus is on the order of 20 to 300 nanometers in diameter. (A nanometer is 1/1000 of a micrometer, which is 1/1,000,000 of a meter.) It's *intracellular* because it lives inside cells. And an *obligate parasite* is one that can only exist through parasitism.

Viruses' inability to reproduce themselves is what makes them dangerous to other organisms. (It also means that, strictly speaking, they're not actually alive themselves.) To reproduce, viruses hijack the machinery of a host cell. They implant their own genetic instructions which turn the cell into a little viral factory whose primary purpose is making copies of the virus—instead of doing whatever else it is the cell would normally do. Most viruses, either quickly or slowly, eventually kill the host cell, which then bursts, spraying copies of the virus around to infect other cells. Other viruses keep the cell alive, but only so that it can pump out a constant stream of new viruses.

Viruses come in a bewildering variety, but in general they share six characteristics:

1. They have only one kind of nucleic acid, either DNA or RNA. (Cells, you'll recall, have both.)
2. They have no protein-synthesizing system (no ribosomes, in other words) and they have no energy-conversion system (no mitochondria or other mechanism to produce ATP).
3. They are not contained by a lipid membrane that they make themselves and they have no internal membranes.
4. They are not affected by antibiotics.
5. They have no cytoskeleton, and they are incapable of movement.
6. They do not grow: Once a virus is formed, it doesn't get any bigger.

The technical term for a fully formed virus, as it exists outside a cell, is a *virion*. A virion consists of little more than genetic material protected by a protein coat called a *capsid*, made up of individual protein sub-units called *capsomeres* (see Fig. 11-1). The nucleic acid and protein coat together are called a *nucleocapsid*.

Viruses have tiny genomes to go along with their tiny sizes. The smallest viral genome is around 500,000 base pairs and the largest is about 5 million base pairs. Viral genomes contain only a few hundred genes which code for enzymes needed for infecting and destroying other cells, polymerases needed for replication, capsid proteins needed for creating new virions, and other proteins that differ from virus to virus depending on their life cycle.

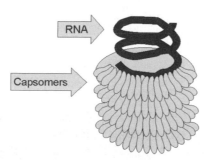

RNA

Capsomers

Fig. 11-1. A section of the tobacco mosaic virus.

Table 11-1. Examples of Viruses Classified by Shape

Nucleic Acid	Capsid Shape	Examples
RNA	Helical symmetry	Tobacco mosaic virus, influenza viruses, potato virus Y
RNA	Cubical symmetry	Polio virus, reoviruses, retroviruses
DNA	Helical symmetry	Smallpox
DNA	Cubical symmetry	Polyoma virus, ΦX174, Shope papilloma virus
DNA	Complex (with head and tail)	T phages, lambda, P22

Viruses are usually symmetrical and can be classified by their shapes (see Table 11-1), which include:

- *Helical symmetry*. The capsid forms a long helix.
- *Cubical symmetry*. The capsid is cube shaped.
- *Icosahedral symmetry*. An icosahedron is a ball with 20 equilateral triangular faces.
- *Complex*. These are viruses with several shapes, such as an icosahedral head and a long helical tail, or an unusual, asymmetrical shape.

Viruses are specialized to attack specific types of cells, and cells that are susceptible to attack have receptors on their surfaces to which the viruses can attack. Cells that don't have these receptors resist attack.

Let's look at the various types of viruses—and how they carry out their genetic attack on cells—in more detail.

Bacteriophages

Viruses that are specialized to attack specific bacterial cells are called bacteriophages, or simply *phages*. (By the way, you only use the plural "phages" when referring to different species of viruses; if you're referring to multiple viruses of the same species, the plural is simply phage. So you could talk about T4 and lambda phages, but you'd say a bacterium was under attack by two lambda phage.)

Many phages are complex, with icosahedral capsids to which a tail is attached. (The viruses that attack eukaryotic cells generally don't have tails.) Most phages have double-stranded DNA as their genetic material, although there are a few that use single-stranded DNA, single-stranded RNA, or double-stranded RNA.

There are two main kinds of phage "life cycles" (which is really a misnomer since viruses aren't actually alive): *lytic* and *lysogenic*.

LYTIC CYCLES

Most phages have only a lytic life cycle (see Fig. 11-2), which kills the host cell in the production of new phage. Phages which kill their host cells are called *virulent*.

The first step of the lytic life cycle is the infection of the host cell, through the adsorption of a virion through a receptor on the surface of the cell. The phage

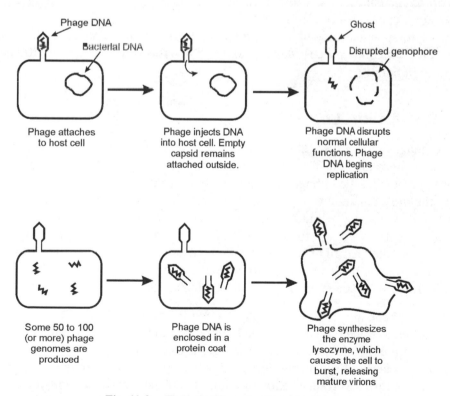

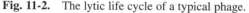

Fig. 11-2.   The lytic life cycle of a typical phage.

then injects its DNA into the host cell. Typically, the protein capsid remains attached to the outside of the bacterium (it's called, appropriately, a *ghost*).

Inside the cell, different phages have different methods of replicating. Most commonly, the host's own RNA polymerase transcribes the phage DNA into mRNAs that are translated into enzymes (needed for replication of the phage genome, transcription, and sometimes to destroy the host cell's DNA), regulatory proteins (which control the timing of the phage genes' activation), and structural proteins (which will form the protein parts of the new copies of the phage).

Construction of the new phage is like an assembly line, with the viral genome being copied by rolling-circle replication and packaged into new protein heads. Details vary from virus to virus; the *E. coli* phage T4, for instance, actually puts more than one full copy of its genome into each head, snipping off the DNA strand once the head is full. This means the order of the genes is different from virus to virus, and there are two *terminal regions* (which makes the genome *terminally redundant*), one at each end of the genome proper, separating it from the extra DNA that was packed in with it. In some other phages, such as the *E. coli* phage lambda, the string of linear DNA being produced by rolling-circle replication is cut in specific places.

Once the phage is assembled, a protein called *lysozyme* is produced that ruptures the host cell (a process called *lysis*) and releases (typically) 50 to 300 new phage in a burst.

LYSOGENIC CYCLE

Some phages have a lysogenic cycle that keeps the host cell alive, at least for a while, to churn out new phage. Because the host cell is not quickly destroyed, as in the lytic cycle, such phages are said to be *temperate* or *nonvirulent*.

In the most common type of lysogenic cycle (see Fig. 11-3), the phage DNA, instead of remaining separate from the host cell's DNA (as in the lytic cycle), is actually incorporated into the host chromosome. In a less common type of lysogenic cycle, the phage DNA doesn't join the host chromosome, but replicates at the same time as the host chromosome does, acting as a sort of plasmid.

The more common type of lysogenic cycle has four main steps:

1. The phage injects its (linear) DNA into the host cell. (The inserted DNA is called a *prophage*.) The injected DNA gets turned into a loop. (Those two terminal regions mentioned earlier are joined up).
2. Some early phage genes are transcribed, producing a repressor protein which turns off transcription of additional phage genes. An enzyme called *integrase* is also produced.

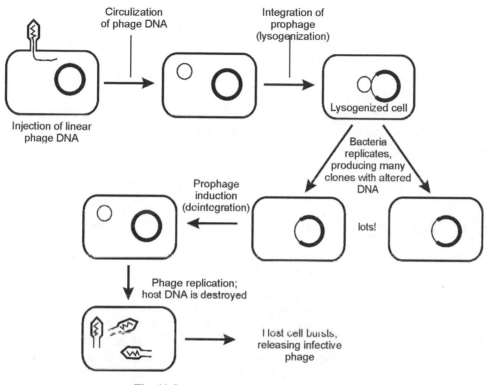

Fig. 11-3.   A typical phage lysogenic cycle.

3. Integrase facilitates the integration of the prophage into a specific site on the host cell's chromosome.

4. As the bacterium replicates, the prophage is replicated along with its own genome.

Some infected cells survive in the lysogenic cycle for a while, then switch to the fatal lytic cycle. One key as to which cycle is triggered seems to be the level of nutrients in the environment. Bacteria become dormant when they run out of nutrients. Since phages can only carry out the lytic cycle in a cell that is actively metabolizing, a starving cell is useless to them. If they can establish a lysogenic cycle in a hungry bacterium, though, they can revive along with the cell when nutrients become available again.

A prophage that has integrated with bacterial DNA can remove itself from the bacterium's chromosome in the event it sustains damage. Part of the bacteria's emergency DNA repair system is a protease enzyme which also has the effect of deactivating the repressor that is keeping the prophage dormant. The prophage then synthesizes an enzyme called *exisionase*, which excises (hence its name) the prophage from the bacterial chromosome. The prophage becomes fully active, and the host cell is plunged into the fatal lytic cycle. This is called *prophage induction*.

In the plasmid form of the lysogenic cycle, once the prophage has been formed into a ring and repressed, it simply lurks in the cell's cytoplasm, replicating once with each replication of the host cell.

Eukaryotic Viruses

There are several differences between eukaryotic viruses and phages, including:

- *Life cycle duration*. The lytic life cycle for a phage typically lasts from 20 to 60 minutes; the life cycle of a eukaryotic virus lasts from six to 48 hours.
- *Number of progeny*. A DNA phage will produce anywhere from 50 to 1000 virions, released when the host cell ruptures. Eukaryotic viruses produce 500 to 100,000 virions per cell, per generation.
- *Efficiency of infection*. Phage progeny are almost 100 percent infectious. Fortunately, most of the progeny of eukaryotic viruses are incapable of causing infection—the number ranges from one in 10 to one in 10,000.
- *Fate of the host cell*. Both the lytic and lysogenic life cycles of phage eventually terminate in the death of the host cell. In eukaryotes, while some host cells die, others continue to grow—and continuously release virions.
- *Impairment of host functions*. Phages usually cause the host cell to quit producing DNA, RNA, and protein very soon after it's infected. Eukaryotic viruses don't usually impair their host cells' function until the late stages of infection.
- *Method of DNA injection*. Phages usually inject their DNA through the cell membrane through their tails, leaving their capsid outside as a ghost. Eukaryotic viruses never inject their DNA; instead, the entire virion is taken inside the cell. (And, as noted earlier, eukaryotic viruses usually don't have tails.)
- *Mutable forms*. There are no known phages that are subject to high rates of mutation. Unfortunately, there are several eukaryotic viruses that exhibit a high rate of mutation, including influenza virus and HIV-1.

ANIMAL VIRUSES

Animal viruses are typically helical or icosahedral, and may be either naked or enveloped. A *naked virus* has a protein capsid, just like a phage. An *enveloped virus* has a capsid, but it also has an envelope made of a portion of the membrane of the host cell, which the virus picked up on its way out.

The viral genome specifies the creation of glycoproteins that are inserted into the membrane. The virion capsid attaches to the ends of the glycoproteins on the cytoplasmic side of the membrane, which in turn causes a bit of the membrane to stick to the virion. In this envelope, the virion is able to pinch off from the cell membrane—a process called *budding*—without leaving a hole.

To infect a cell, the virion attaches to a specific receptor on the cell membrane which either fits the capsid of a naked virus or the glycoproteins in the envelope of an enveloped virus. Once inside the cell, the capsid and envelope (if any) are removed, releasing the viral genome, which may be DNA or RNA, single-stranded or double-stranded, linear or circular (if DNA—there are no known circular RNA viral genomes). DNA viruses are generally replicated by the host cell's chromosome, while RNA viruses remain in the cytoplasm.

Viruses cause four main types of infection in animals:

1. *Acute or lytic.* In this kind of infection, viruses follow the lytic life cycle (described earlier in the section on phages) and quickly kill their host cells by causing them to rupture and release progeny virions.
2. *Latent.* This corresponds to the lysogenic life cycle in phages: The virus infects the cell, but remains dormant until specific conditions are met.
3. *Persistent.* In this kind of infection, new virions are slowly released from the cell surface, but the cell isn't killed. This produces the enveloped viruses just described.
4. *Transformative.* This kind of infection both produces new virions and transforms a normal cell into a cancerous cell by inserting an oncogene (as described in Chapter 8) into the cell's chromosome.

DNA and RNA viruses follow different paths to replication, transcription, and translation when they infect animal cells.

A typical double-stranded DNA virus attaches to the surface of the cell, is taken inside, and then loses its capsid (a process called *uncoating*). The host cell's own enzymes replicate the viral DNA and transcribe it into mRNA, which the host's ribosomes translate into viral capsid proteins or (sometimes) enzymes that favor replication of the viral DNA over replication of the host cell's own DNA. The capsid proteins, or capsomeres, form capsids around the replicated viral DNA, and then are released either through the rupturing of the host cell or by budding (which produces the enveloped virions described earlier). A single-stranded DNA virus follows the same general path, except its single strand of DNA is first matched up with nucleotides from the cell to become double-stranded DNA, after which transcription and translation follow.

RNA virus life cycles come in more flavors than DNA life cycles. Most host cells can't replicate or repair RNA because they don't have the right enzymes. That makes RNA viruses more subject to mutation, and also means that the RNA virus genome has to include genes to produce the enzymes necessary for its own replication—or bring the enzymes themselves along for the ride when they infect a new cell.

Single-stranded RNA genomes are labeled either (+) or (−). A *(+)RNA strand* acts as an mRNA within the cell, coding for (at a minimum) capsid proteins and the enzymes needed for RNA replication. A *(−)RNA strand* is complementary to the mRNA strand needed to code for those things, and so has to bring with it an enzyme that can synthesize a (+) strand from a (−) strand—after which synthesis of proteins and enzymes can proceed as needed.

Double-stranded RNA genomes are replicated more or less like double-stranded DNA ones, using a polymerase called *RNA replicase*. Finally, retroviruses carry with them reverse transcriptase, an enzyme that copies their RNA genome into DNA which can be integrated into the genome of the host cell or used for transcription. As noted in Chapter 8, some retroviruses carry oncogenes in this fashion that turn cells cancerous. Another example of a retrovirus that inserts dangerous genes is HIV, the AIDS virus. It's the most complex retrovirus known because it contains at least six extra genes.

PLANT VIRUSES

Plant viruses can be rods or polyhedrons, and generally have (+)RNA genomes. Interestingly, some plant viruses can't replicate unless there are two different virions inside the host cell, each bearing part of the genome. These kinds of viruses are said to have *segmented genomes*.

There are some DNA plant viruses, such as the cauliflower mosaic virus, which has a double-stranded DNA genome in a polyhedral capsid, and the *geminiviruses*, which have a pair of connected capsids, each of which contains a circular, single-stranded DNA molecule about 2500 nucleotides long. In some geminiviruses the paired molecules are identical, while in some they are very different.

Finally, there are *viroids*. These are tiny circular single-stranded RNA genomes of just 270 to 380 nucleotides, too few to code for any proteins. Just as viruses are hundreds or thousands of times smaller than the cells they infect, viroids are thousands of times smaller than viruses. They rely on the host cell's enzymes to replicate themselves. Despite their small size, they cause plant diseases because some of the segments in their genome confuse the cell's translation mechanism.

The ability of viruses to hijack the normal functions of cells and insert their own genetic material into them had made them extremely valuable in genetic engineering, one of the most fascinating and controversial areas of research in the field of genetics—and the subject of our next chapter.

Quiz

1. The first virus recognized as such was
   (a) smallpox.
   (b) influenza.
   (c) tobacco mosaic virus.
   (d) potato blight.

2. Viruses are
   (a) symbiotes.
   (b) parasites.
   (c) tiny bacteria.
   (d) mutated algae.

3. A capsid is
   (a) the protein coat of a virus.
   (b) a tight cap worn by viral researchers.
   (c) a European flower often infected by a viral disease.
   (d) a cell that's been infected by a virus.

4. Viruses that attack bacteria are called
   (a) prokaryoviruses.
   (b) plasmiditrons.
   (c) bacteriophages.
   (d) bacteriofleas.

5. For the host cell, the lytic cycle ends when
   (a) it replicates.
   (b) it bursts open.
   (c) it collapses.
   (d) it becomes cancerous.

6. The phage life cycle which does not kill the host cell is called
   (a) the merciful cycle.
   (b) the nonlytic cycle.
   (c) the semiphage cycle.
   (d) the lysogenic cycle.

7. The life cycle of eukaryotic viruses is
   (a) shorter than that of phages.
   (b) about the same as that of phages.
   (c) longer than that of phages.
   (d) longer for plants, shorter for animals.

8. An enveloped virus has
   (a) a lipid membrane picked up from the host cell.
   (b) a very thick capsid.
   (c) a very long tail.
   (d) been swallowed by a white blood cell.

9. HIV and some other retroviruses have
   (a) fewer genes than other viruses.
   (b) no protein coat.
   (c) extra genes not required for their own replication.
   (d) DNA that reads in the opposite direction to other viruses.

10. Viroids are
    (a) extremely tiny RNA genomes.
    (b) dead viruses.
    (c) artificial viruses constructed using nanotechnology.
    (d) glands in the throat that often become inflamed.

Genetic Engineering— Sculpting the Code

In November of 1972, in a delicatessen in Hawaii, two molecular biologists, Stanley Cohen of Stanford University and Herbert Boyer of the University of California at San Francisco, chatted over a snack after a long day of meetings at a scientific conference.

As they talked about their respective work, they realized that they each had knowledge, tools, and techniques that the other needed. By the time they finished their corned beef sandwiches, they had planned a series of experiments that would revolutionize genetics.

Cohen was interested in placing new genetic material into bacteria by removing the plasmids from one cell and inserting them into another. Nature does that through conjugation, but Cohen had found ways to do it artificially.

Boyer was working on something quite different.

Restriction Enzymes

Nucleases are enzymes that can break the bonds holding nucleotides together. *Deoxyribonucleases* attack DNA molecules, while *ribonucleases* attack RNA molecules (especially where they are single-stranded). Some nucleases remove terminal nucleotides one at a time; these are called *exonases*. Those that break the backbone of DNA or RNA molecules somewhere other than at the end—splitting the strand—are called *endonucleases*.

Boyer's work was focused on a special kind of nuclease called a *restriction endonuclease*—in particular, a kind labeled Type II (there are also Type I and Type III). Type II restriction enzymes recognize and bind to a specific double-stranded DNA sequence called a *restriction site,* and break the backbone of each strand somewhere within 20 base pairs of that site (see Fig. 12-1).

The restriction site typically consists of four to eight base pairs, which may either be continuous (for example, GAATTC) or interrupted (GTXXAC, where X can be any base). They're symmetrical, which is to say they read the same (but in the opposite direction) on the other strand of DNA. In the examples above, the opposite strand would code CTTAAG and CAXXTG. This symmetry makes the complementary sequences a *palindrome.*

Restriction enzymes occur naturally in bacteria, where they help fend off infection by viruses. The recognition sites in the bacterial DNA are modified by the addition of a methyl group ($-CH_3$), a process called *methylation* that is carried out by enzymes called (what else?) *methylases.* This keeps the restriction enzymes from cleaving the genome at those points. An invading virus, however, will have unmodified recognition sites, and can be degraded and possibly destroyed by the bacteria's restriction enzymes. This is called the *restriction and modification system.*

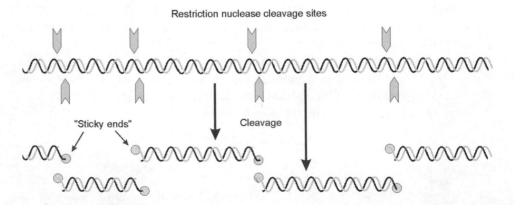

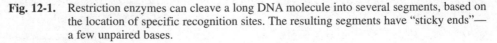

Fig. 12-1.   Restriction enzymes can cleave a long DNA molecule into several segments, based on the location of specific recognition sites. The resulting segments have "sticky ends"—a few unpaired bases.

Naming Restriction Enzymes

Restriction enzymes are named after the species of bacteria from which they come. Typically, the name begins with three letters. The first one, capitalized, is the first letter of the genus of the bacteria. The subsequent letters are the first two letters of the species. These three letters are all italicized (because scientific names are traditionally italicized). Following the three letters are Roman numerals, non-italicized, to indicate which of possibly several enzymes were isolated from that particular species of bacteria.

If the enzyme is from a specific strain of a particular species, a fourth letter to indicate that strain will follow the name, and is not italicized.

Some examples:

- *Pst*II — The second enzyme isolated from *Providentia stuartii*
- *Eco*RI — The first enzyme isolated from *E. coli* strain RY13
- *Hind*III — The third enzyme isolated from *Haemophilus influenzae* strain Rd

Restriction enzymes attracted the attention of Boyer and other scientists because they offered a method of splitting extremely long DNA molecules into manageable chunks, and doing so in a predictable manner. Restriction enzymes also do something else very useful: They leave an incomplete sequence of unpaired bases at the end of each fragment. For example, Boyer's particular focus of interest, the enzyme *Eco*RI, always left the strand TTAA at one end and its mirror image, AATT, at the other.

Any fragment of DNA cut by a particular restriction enzyme can thus be attached to any other fragment of DNA cut by that same restriction enzyme, because the two fragments have complementary "sticky ends." The enzyme ligase can be used to seal the fragments together.

During their delicatessen discussion in Honolulu, Boyers and Cohen realized that by joining forces they could do something no one had ever done before—directly engineer an organism's genome.

The First Experiments

Boyer and Cohen began their joint experiments with coworkers Annie Chang and Robert Helling in California in the spring of 1973. They used *Eco*RI to cleave the DNA in plasmids from two different strains of *E. coli*. One plasmid contained a gene that provided resistance to the antibiotic tetracycline, while the other had a gene that provided resistance to the antibiotic kanamycin. Once the plasmids' ring-shaped genomes had been split by the enzyme, the researchers joined them back together to form one large plasmid genome containing both antibiotic-resistance genes.

They put the new plasmid into a strain of *E. coli* that wasn't resistant to either type of antibiotic, then transferred the bacteria to a culture dish containing both drugs. Some of the bacteria survived—which meant their new genes were fully functional.

Next, Boyer and Cohen combined plasmids from two different kinds of bacteria; then they put a gene from a frog into a plasmid. In both cases, the genes were expressed in the bacteria that received the new plasmids, and passed down to the bacteria's progeny.

The scientists had created entirely new organisms by the direct manipulation of genes. The era of genetic engineering had begun.

Amplifying DNA

To introduce new genes into an organism, you first have to have a ready supply of genes. And so, the first step of any genetic engineering experiment is the isolation and cloning of genetic material.

GENE CLONING

In popular usage, a clone is thought of as an identical copy of a higher organism, like the late, famous Dolly the sheep. To molecular biologists, however, a clone is a population of genetically identical (except for mutations which arise during the cloning process) organisms, cells, viruses, or DNA molecules, all derived from the duplication of a single originating organism, cell, virus, or DNA molecule.

To clone a virus, you only have to infect a single cell with a single virion. All the virions produced by the infected cell will be clones of the original. Since these clones can then go on to infect new cells, you can produce a lot of copies of that initial virion very quickly. They show up as isolated clear patches (*plaques*), caused by the death of infected cells, on a layer (*lawn*) of uninfected bacteria in a Petri dish.

You can clone a cell simply by placing it in isolation in a growth medium (typically *agar*, a substance made from seaweed). Bacterial and mammalian cells can be easily cloned in this manner.

However, the goal of cloning, from the point of view of DNA research, is not to create a lot of identical copies of a normal cell, but to obtain a sizeable number of copies of a particular DNA fragment.

And in the case of genetic engineering, the goal of cloning is most often to create a sizeable number of copies of a specific gene that does something you'd like to get a genetically modified organism to also do.

Before you can make multiple copies of a gene, you must first isolate it from the rest of the genome. The first step in that process has traditionally been to create a *library* that breaks the genome down into smaller chunks. If you're lucky, one of those chunks will contain the gene you're looking for.

A library, in this usage, has nothing to do with books. Rather, it's a collection of DNA fragments that all come from the same source—a collection of clones.

To create a library, you need a *vector*—something that can insert new DNA into a host cell and get the host cell to replicate it. One often-used vector is a plasmid (we discussed plasmids back in Chapter 9); phages (discussed in Chapter 11) can also be used.

Choosing a Cloning Vector

Not just any old virus or plasmid can be used as a vector. An ideal vector for genetic engineering has three characteristics:

1. It needs lots of cloning power. It needs to be able to make many replicas of itself within the host cell.
2. Its genome should have a single recognition site for several restriction enzymes. That way, the foreign gene can only be inserted in a single location.
3. It should do something to its host cell—make it antibiotic-resistant, for instance—that will set that cell apart from non-engineered cells, so the inserted gene can be easily isolated.

DNA from the donor organism is cut into many pieces using a restriction enzyme that also cleaves the DNA of the plasmid (the vector in this example) in a single location. The bits of genomic DNA and the cloven plasmid genomes are mixed; their "sticky" ends join up with one another, pretty much at random. The result is many different plasmids, each containing a different bit of the donor DNA, which collectively (hopefully) contain all of it.

These plasmids are inserted into recipient bacterial cells (made permeable to the plasmid's naked DNA by treatment with a cold calcium chloride solution).

The transformed bacterial cells are spread thinly on agar so that each cell is free to reproduce in isolation from the others.

The plasmid used as a vector usually also contains genes for two kinds of antibiotic resistance. Researchers use a restriction enzyme to cleave the plasmid DNA in a place that disrupts one of those genes. Any cell that did not take in a plasmid is killed by one of the antibiotics, while any cell that took in a plasmid that did not receive a segment of the DNA being cloned will thrive happily in agar containing either antibiotic. The cells containing plasmids that did take up the foreign DNA, however, will only grow in agar that contains the antibiotic whose resistance gene was not disrupted by the restriction enzyme (see Fig. 12-2). Those foreign DNA-containing cells multiply, forming colonies, or clones. All the cells in each colony contain a copy or multiple copies of the same fragment of the original DNA, but each colony contains copies of a different fragment.

Similar techniques can be used with viral or phage vectors.

Libraries, once created, can be stored as permanent reference resources (which is why they're called libraries, presumably).

Libraries are often used in the effort to isolate specific genes. There are two ways to find the specific clone within the library that contains a DNA fragment that includes a gene of interest. One is to look for a specific sequence of base pairs (if known). The other is to look for the specific protein encoded by the gene. In either case, the process is called *screening*. There are many techniques, depending on the vector used and the gene being sought.

One example of a screening process, where the DNA sequence being sought is known, is illustrated in Fig. 12-3.

Bacterial colonies or phage plaques are transferred from an agar-filled Petri dish to a solid disk, usually made of nitrocellulose, by the simple method of laying the disk on the agar, then lifting it up again. This doesn't remove the colonies from the agar, but it does transfer enough material to the nitrocellulose for analysis.

Next, the DNA on the nitrocellulose disk is unwound by immersing the disk in a chemical solution. Then the disk is placed in a solution that contains a single strand of DNA or RNA tagged with a radioactive atom complementary to one of the strands in the clone. This tagged single strand is called a *probe*.

Wherever the probe joins up with one of the recombinant strands, it leaves a radioactive spot on the disk which can be registered on photographic film, and directly correlated to a location on the original agar disk.

If the DNA sequence of the gene being sought is unknown, but a protein the gene codes for is known, another screening method is to create clones using vectors that allow the genes they carry to be expressed in their new host cells. The

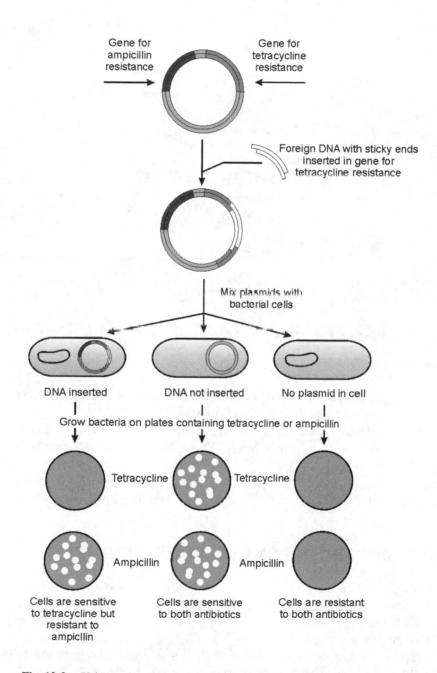

Fig. 12-2. Using a plasmid vector with two antibiotic-resistance genes makes creation of a library easier.

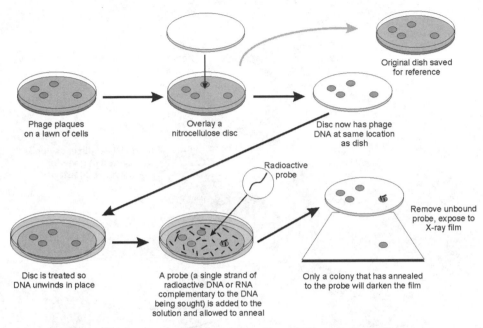

Fig. 12-3. Here's one method of screening clones for a particular DNA sequence.

cells begin producing the protein the gene encodes, and researchers can then search the library for the protein in question.

Once a desired clone is located, it can be picked off the agar plate with a needle and allowed to multiply freely. The recombinant DNA can be chemically purified from the cells for use in the laboratory, and the clone that produces it can be stored and regrown as needed. This provides genetic engineers with an endless supply of a particular gene for insertion into other organisms.

PCR: AMPLIFYING DNA WITHOUT CLONING

Since 1985, geneticists have had a remarkable new way to make copies of DNA that doesn't involve cloning: *polymerase chain reaction,* or *PCR,* a powerful process that can make millions or billions of copies of any selected sequence of DNA in just a few hours.

The PCR technique uses *primers,* short sequences of synthesized DNA, typically 12 to 20 nucleotides long, designed to match up with the nucleotides on opposite strands at each end of the region to be copied.

The template—the DNA sequence to be duplicated—doesn't have to be highly purified, and a large number of copies of it aren't required.

Once the template is selected and the primers prepared, PCR proceeds in three steps (see Fig. 12-4):

1. *Denaturation.* The DNA containing the section to be copied is heated, which causes the complementary DNA chains to separate. Then the primers are added to the mixture, along with a heat-tolerant DNA polymerase.

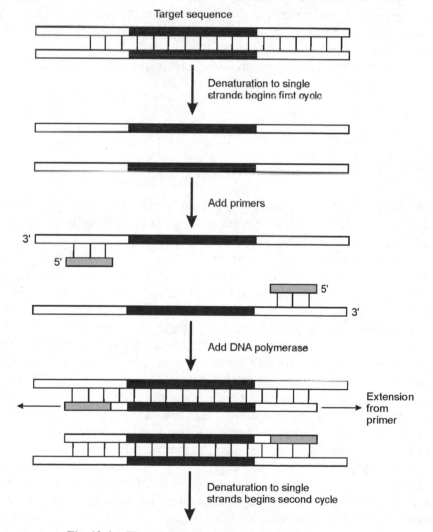

Fig. 12-4. The polymerase chain reaction (PCR).

2. *Annealing.* As the mixture cools, the primers bind to the single-stranded chains.
3. *Extension.* The polymerase extends the primer through the rest of the DNA fragment, creating double-stranded DNA molecules again.

These three steps are repeated over and over. During each cooling phase any excess primers (or newly added primers) bind to more template strands and are again extended, producing more and more double-stranded DNA molecules.

Each cycle takes several minutes, but in just a few hours—after 20 cycles—a single DNA molecule can be copied about one million times. And after just 30 cycles, there are around one billion copies—plenty for researchers to work with.

As with DNA copied through cloning techniques, probes can be used to identify specific genes or sequences of interest.

Modifying Proteins

The first tangible result of the genetic engineering breakthrough was the ability to produce medically and economically important proteins using genetically modified *E. coli* bacteria.

The process seemed like it should be straightforward: Clone the gene that produces the protein wanted—insulin, say, or human growth hormone—insert it into a plasmid vector, then insert the plasmid into *E. coli.* The bacteria would begin producing insulin. Since it could be grown quickly in large vats, significant quantities of the protein would be readily recoverable.

It turned out to be trickier than that. Prokaryotes like *E. coli* do not have introns like the genomes that ordinarily code for hormones. That meant they couldn't produce the correct mRNA to code for the hormones. As well, many translation products of eukaryotic genes require additional chemical interactions before they become active—and *E. coli,* not surprisingly, doesn't provide those interactions. It also turned out that many eukaryotic proteins are actually toxic to *E. coli,* while others are degraded by the bacteria's enzymes.

Fairly quickly, however, these problems were worked out. Today, a number of useful proteins are produced using several different hosts: not just *E. coli,* but *Bacillus,* yeasts, and other fungi like *Aspergillus* and *Fusarium,* and even plant, mammalian, and insect cell cultures.

Once a stable culture of cells containing the successful genetic modification has been created, more of the cells can be grown as required to produce the sought-after protein. Among the proteins produced in this way are human insulin (for use by diabetics), blood-clotting factors (required by hemophiliacs), and human growth factor. Some vaccines are also produced in this fashion, such as the one for hepatitis B. There are dozens more examples.

Genetic Engineering of Animals

There are two ways to modify animals genetically: by altering *somatic cells* (body cells), and by altering *germ cells*. The former only affects the individual organism receiving the modification, while the latter introduces new, inheritable traits.

Modifying somatic cells involves four steps:

1. Remove cells from the organism.
2. Culture them.
3. Transform them with a vector containing the gene you want to introduce.
4. Reintroduce the transformed cells into the organism.

This is the basic template for gene therapy in humans, which we'll look at in more detail later in this chapter.

Altering germ cells produces organisms with new, hereditable traits. An organism containing new genes that it passes on to its offspring is called *transgenic;* the newly introduced gene is called a *transgene.*

The process of producing transgenic animals is more complex than that of inserting genes into bacteria, as you might expect. Nevertheless, many animals have been engineered to date. Here are just a few examples:

- Mice have been engineered in a number of ways. Some have been given human genes to provide a non-human model of human diseases. For example, mice have been genetically engineered that express the human gene for the polio virus receptor; unlike normal mice, they can be infected by the polio virus and even develop the disease's symptoms. Another example is mice with a human immune system, which enables researchers to learn more about the immune system without using human patients.

- Sheep have been engineered to produce the human gene for alpha1-antitrypsin in their milk. People who inherit two non-functioning genes for this protein develop a disease called Alpha1-Antitrypsin Deficiency, which damages the lungs and sometimes the liver.
- Chickens have been successfully engineered to produce potentially valuable human proteins in their egg whites.
- Researchers in Guelph, Ontario, have created the Enviropig, genetically engineered to produce 60 percent less phosphorus in its manure.
- Goats have been genetically engineered to produce spider silk proteins and human insulin in their milk (not at the same time).

The precise method used varies from animal to animal, but two methods used in mice provide a sense of what is involved (see Fig. 12-5).

The *embryonic stem cell method* makes use of embryonic stem cells—cells which have the potential to become any kind of tissue—harvested from the inner cell mass of mouse *blastocysts* (mice in the very early stages of development after conception).

The gene researchers want the mouse to express is isolated and cloned using the methods described earlier and, using an appropriate vector, inserted into the stem cells, along with some necessary promoter and enhancer sequences. The resulting cells are screened to find those in which the gene has been both successfully inserted and inserted into the correct place in the genome. The successfully transformed stem cells are cultured, then injected into the inner masses of mouse blastocysts, which are then transferred into a *pseudopregnant* female—one who has been mated to a sterile male to trigger the hormonal changes required to make her uterus receptive to the embryos.

No more than one-third of the embryos will develop into healthy pups. Of those, no more than 10 to 20 percent will have the desired gene, and they'll be heterozygous, having only a single copy.

The pups are mated with each other, and the resulting offspring are screened to find the one in four (as per Mendel's Laws) that are homozygous, with two copies of the gene. Breeding the homozygous mice together establishes a new strain of transgenic mice, all of whom express the gene.

The second method is the *pronucleus method*. In it, freshly fertilized eggs are harvested, and the donor DNA (prepared as in the previous method) is injected into the head of the sperm cell before it forms a pronucleus. Once the fertilized egg has formed a nucleus and divided into a two-cell embryo, it's implanted into a pseudopregnant female. After that, everything proceeds as in the first method.

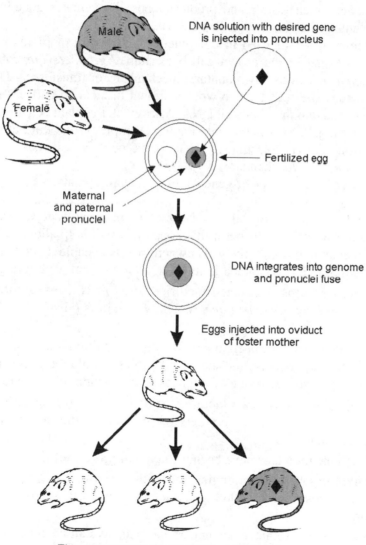

Fig. 12-5. Creating a transgenic mouse.

Genetic Engineering of Plants

In plants, unlike in animals, there aren't any real distinctions between somatic cells and germ cells, because the somatic tissues of plants can be grown into mature plants which flower and produce seeds. That makes genetically engineering plants somewhat easier than animals.

One common vector used in the genetic engineering of plants is a family of plasmids found in the crown-gall bacterium, *Agrobacterium tumefaciens*. *A. tumefaciens* plasmids have a natural mechanism for transferring DNA into a plant. In nature, the cells bind to wounded plant tissue, and part of the plasmid DNA is inserted into the plant cell DNA. As a result, the plant cells form a tumor called a crown gall. This is the only known example of this kind of transfer of genes from a bacterial plasmid to a eukaryotic cell—and it makes a splendid genetic engineering mechanism.

Here are some examples of genetically engineered plants:

- Rice has been modified to manufacture beta-carotene, a precursor of Vitamin A, in its endosperm. It's hoped this will help alleviate Vitamin A deficiency in much of the world where rice is a staple. (The beta carotene in normal rice is contained in its husk, which is removed during milling.)

- Numerous crops have been modified to express a toxin normally produced by *Bacillus thuringiensis*. This *Bt* toxin is poisonous to a number of insect pests.

- Tobacco plants have been modified with the addition of a gene that confers resistance to tobacco mosaic virus (which would probably please our friend Martinus Beijerinck from back at the beginning of Chapter 11).

- Canola plants have been modified to be resistant to a specific herbicide, so that the herbicide can be used to kill weeds in the field without harming the plant.

- Transgenic tomatoes have been created that grow well in saline soils.

- Some crops have been engineered with genes that make them produce sterile seeds, thereby forcing farmers to buy fresh seeds the following year.

- Corn, tobacco, tomatoes, potatoes, and rice have all been modified to produce a variety of therapeutic proteins, including human growth hormone (the gene is inserted into the chloroplast DNA of tobacco plants), humanized antibodies, protein antigens for use in vaccines, and more.

Gene Therapy

Gene therapy is the genetic engineering of human somatic cells in an attempt to alleviate a disease with a genetic component.

In 1990, a three-year-old girl named Ashanthi deSilva became the first person to be treated via gene therapy. Ashanthi suffered from *ADA deficiency*. The cells of people with ADA deficiency can't make an enzyme called *adenosine deaminase,* which normally breaks down certain substances in the cells. Although these accumulated substances don't harm most cells, they're toxic to some of the white blood cells that play a key role in the human immune system. As a result, children with ADA deficiency have little immunity and are almost constantly sick. Most die before they are two years old.

In an attempt to treat the disease genetically, a doctor named W. French Anderson and his colleagues removed some of her blood, filtered out the white blood cells, and mixed those cells with an altered retrovirus containing the ADA gene. The cells were allowed to multiply for 10 days, and were then transfused back into Ashanthi (see Fig. 12-6).

At first the treatment was repeated monthly. Ashanthi also took regular shots of a PEG-ADA, a form of ADA that doesn't break down in the blood the way regular ADA would. After a few years, however, the treatment was reduced to just once a year and her dose of PEG-ADA was cut in half. Her immune system improved steadily and she was able to begin living the life of a normal child.

Besides retroviruses, other vectors that have been used in gene therapy experiments include *adenoviruses* (double-stranded DNA viruses that cause the common cold, among other ailments), *adeno-associated viruses* (single-stranded DNA viruses that insert their genetic material at a specific site on chromosome 19), and *Herpes simplex* virus (double-stranded DNA viruses that infect neurons specifically. The virus that causes cold sores is *Herpes simplex* type I.)

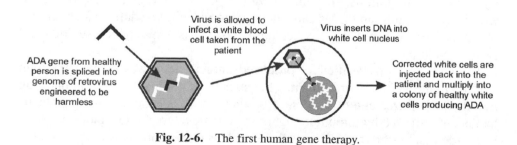

Fig. 12-6.   The first human gene therapy.

Other approaches have included introducing therapeutic DNA directly into target cells and creating an artificial lipid sphere with a liquid core containing the therapeutic DNA. These *liposomes* can pass through target cells' membranes.

Researchers are even experimenting with creating an artificial 47th chromosome that could carry substantial amounts of genetic code. However, delivering such a large molecule into the nucleus of the target cell is likely to be difficult.

Despite its early success, gene therapy has yet to live up to its promise.

In 1999, a major setback was the death of Jesse Gelsinger, an 18-year-old participant in a gene therapy trial who apparently had a severe immune response to the adenovirus vector. In 2002, in another setback, two children in France developed a leukemia-like condition while being treated for an immunodeficiency disease via gene therapy that used retroviruses to insert genes into blood stem cells.

The body's immune response to invaders both raises the risk of a severe immune reaction to gene therapy treatments and makes it difficult for those treatments to get a foothold. Other problems gene therapy has yet to overcome include the difficulty of getting the therapeutic DNA to remain functional long enough to benefit the patient, the risks posed by using viruses as vectors, and the fact that many diseases involve the action of more than one gene—making them difficult targets for gene therapy.

But research continues and advances are being made.

Cloning

A clone of a higher organism is just like a clone of a gene—a genetically identical copy. Plants have been cloned for thousands of years: The plant produced from a leaf cutting is genetically identical to the plant from which the cutting came.

There are also naturally occurring clones in the animal kingdom. With the right chemical stimulus, the unfertilized eggs of some small invertebrates, worms, fish, lizards, and frogs can develop into full-grown adults that are clones of the mother. (This process is called parthenogenesis.) And, of course, identical twins are clones of each other.

However, the first artificially produced animal clone was created by John Gurdon in the 1970s. He transplanted the nucleus of a somatic cell from one frog into the unfertilized egg of a second frog that had been *enucleated*—its nucleus had been destroyed by ultraviolet light. The egg developed into a tadpole genetically identical to the frog whose cell the nucleus came from. However, it did not grow into an adult frog.

This type of cloning is called *somatic nuclear transfer*. It was not successfully carried out in mammals until Ian Wilmut and colleagues at the Roslin Institute in Edinburgh successfully cloned a sheep named Dolly in 1997. Wilmut took a nucleus from a mammary gland cell of a Finn Dorsett sheep and transplanted it into the enucleated egg of a Scottish blackface ewe. The nucleus and egg were forced to fuse (and stimulated into dividing) with a jolt of electricity, a process called *electrofusion*. The new cell divided and was placed in the uterus of a blackface sheep to develop (see Fig. 12-7). Dolly was born some months later. The process had been tried 275 times before it was successful.

Since Dolly, many other mammals have been successfully cloned, including cattle, goats, mice, pigs, and cats.

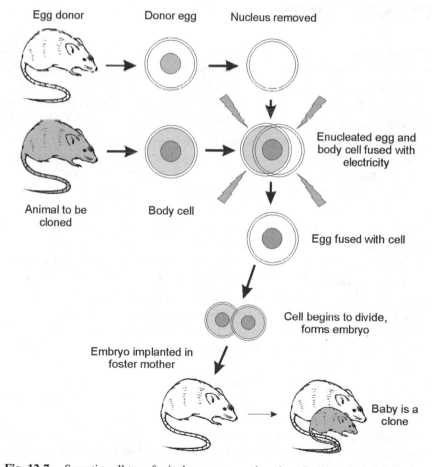

Fig. 12-7. Somatic cell transfer is the process used to clone Dolly and other animals.

Genetic engineering techniques can be combined with cloning techniques. A recombinant somatic cell nucleus can be implanted in the enucleated egg, resulting in an animal with genes designed to code for specific proteins. One goal is to create animals that produce therapeutic human proteins. (Transgenic sheep have been created that produce human insulin.) Another use might be to create animals with human-compatible tissues and organs for therapeutic or research purposes. Cloning would be the preferred method of maintaining a line of such modified animals to ensure that the desired traits weren't bred out by the normal genetic shuffling of sex.

Concerns and Controversy

After Boyer and Cohen announced their development of recombinant DNA technology, geneticists were excited by the new vistas of scientific exploration that it opened. But it also raised red flags in the minds of some scientists, including Paul Berg, a colleague of Cohen at Stanford who had conducted his own experiments with moving genes from one organism to another. He had transferred (without benefit of restriction enzymes, which made it much harder) a gene from SV40, a virus that infects monkeys, into the genome of the phage lambda.

He didn't put the resulting DNA into *E. coli,* however, or show that it would still function there—in part because of concerns raised by Robert Pollack, a geneticist working at Cold Spring Harbor. Pollack worried about putting a virus that caused cancer in mice and hamsters (though harmless in monkeys) into a bacteria that happily lives in the human gut.

Late in 1973, Berg and 77 other molecular biologists wrote to the magazine *Science* and recommended that the National Institutes of Health (NIH) establish safety guidelines for the kinds of experiments being carried out by Boyer and Cohen and, by that time, many others. A few months later, in July of 1974, in a second letter, the concerned scientists asked for a moratorium on gene splicing until the possible hazards could be evaluated and any necessary safety precautions could be drawn up.

In February of 1975, 140 molecular biologists met in Asilomar, California, and concluded that recombinant DNA work should proceed, but only with the provision that appropriate safeguards were put in place to ensure that newly created organisms did not escape the laboratory. In 1976, the NIH drew up guidelines very similar to those created by the scientists at the Asilomar meeting, that it required all scientists receiving federal funding to follow. The NIH also created a Recombinant DNA Advisory Committee (RAC) to approve new genetic engineering experiments. Many of those early concerns have been alleviated, but the RAC still exists.

That was the beginning of ongoing controversy over the safety and ethics of genetic engineering.

Some of the concerns expressed include:

- *Health-related issues.* These usually center around genetically modified crops. Could they be toxic? Might they be less nutritious? Could they trigger food allergies?
- *Environmental and ecological issues.* Does the use of genetically engineered crops lead to more (or less) use of agricultural pesticides? Are plants modified to be hardier than normal likely to escape from cultivated areas and become superweeds? Can the new genes added to genetically engineered plants somehow transmit their genes to other crops or weed species? Do plants modified with the Bt gene (which produces the protein toxic to insects mentioned earlier) hurt non-target insects?
- *Economic and ethical issues.* Should new organisms produced through genetic modification be patentable? Are crops modified to produce sterile seed ethical? Do clones threaten biodiversity? Is human cloning ever ethical—even to produce stem cells with possible medical value? And who decides?

It's safe to say, at this point, that while we certainly have the capability of moving genes from organism to organism, more research remains to be done on when and where it is either safe or ethical to do so.

Quiz

1. Enzymes that can break the bonds that hold the DNA backbones together are called
   (a) nucleases.
   (b) fissionases.
   (c) backbreakers.
   (d) debasers.

2. "Sticky end" refers to
   (a) a cellular adhesive that holds cells together to form tissues.
   (b) a short, complementary, single-stranded DNA segment at each end of a DNA fragment.
   (c) geneticist slang for a laboratory accident that spoils a DNA sample.
   (d) the tail of a phage used as a genetic engineering vector.

3. The first genetically engineered organism was
   (a) a sheep.
   (b) yeast.
   (c) the *Haemophilus influenzae* Rd virus.
   (d) *E. coli.*

3. A clone is
   (a) a genetically identical copy of another organism.
   (b) a genetically modified copy of another organism.
   (c) a genetic mixture of two different organisms.
   (d) a bacterium modified to produce a human hormone.

4. PCR (polymerase chain reaction) is used to
   (a) grow *E. coli* in the laboratory.
   (b) power cell activity.
   (c) make many copies of a DNA sequence quickly.
   (d) clean dried DNA from laboratory glassware.

5. Somatic cells are
   (a) body cells.
   (b) dormant cells.
   (c) phage-infected *E. coli* cells.
   (d) cells that can't be cultured.

6. A genetically engineered organism that can pass on its new genes to its offspring is said to be
   (a) photogenic.
   (b) transgenic.
   (c) hygienic.
   (d) bionic.

7. The first disease treated by gene therapy was
    (a) cancer.
    (b) cardiovascular disease.
    (c) an immunodeficiency disease.
    (d) hypothyroidism.

8. Somatic nuclear transfer is used to
    (a) move radioactively tagged DNA molecules from one cell to another.
    (b) treat diabetes.
    (c) rapidly duplicate RNA molecules.
    (d) clone animals.

9. The first mammal to be successfully cloned was a
    (a) cow.
    (b) sheep.
    (c) pig.
    (d) human.

Evolution—Change Driven By Genetics

As noted in Chapter 1, when Charles Darwin published his 1859 book *On the Origin of Species*, one of the charges critics leveled was that he didn't have a satisfactory explanation for how offspring inherited characteristics from their parents. Darwin had to admit they were right—Mendel's work, and everything that followed from it, was still in the future.

But once the principles of genetics were developed, they provided both the mechanism for evolution and new data to advance its study.

What Is Evolution?

Evolution is change over time. Biologically, it refers to heritable changes in a population, spread over many generations. Another definition you'll sometimes see is "any change in the frequency of alleles within a gene pool from one generation to the next" (which will make more sense when you get to the section on population genetics later in this chapter).

Given enough time, this slow change over time creates entirely new organisms.

Darwin called the driving force of evolution *natural selection.* It can also be boiled down to four basic tenants:

1. Each generation, more individuals are born than can survive.
2. The traits exhibited by individuals vary, and those variations can be passed on to the next generation.
3. Those individuals with traits best suited to the environment are more likely to survive.
4. When a breeding population is isolated from other members of its species, new species will form.

Evidence of this evolution of species can be observed in the fossil record. Scientists believe the history of life looks something like this (in, to say the least, highly abbreviated form):

- *4.5 billion years ago.* The Earth forms.
- *4.2 billion years ago.* Oceans cover the surface.
- *3.7 billion years ago.* Rudimentary cells form with the ability to reproduce and transmit genetic information. (RNA is thought to have been the first genetic information molecule. These ancestor cells gave rise to the three domains of organisms alive today: the Bacteria, the Archaea, and the Eukarya.)
- *0.7 billion years ago.* The first multicellular organisms, called metazoans, appear and rapidly expand. (Metazoans showed a tremendous range of body plans, many of which have vanished due to the extinctions of major groups.)
- *5–6 million years ago.* The divergence of our ancestors from our most recent common ancestor with chimpanzees, our closest living animal relatives.

Population Genetics

Evolution is less concerned with genetics as it affects individuals than with genetics as it affects populations, a branch of the science called, logically enough, *population genetics.*

Population has a special meaning within population genetics. It's not just any group of individuals. Rather, a *Mendelian population* is a group of sexually reproducing organisms that have a relatively close degree of genetic relationship, live within a defined geographic area, and interbreed.

The *gene pool* is a hypothetical mixture of all the genes contained within the population. It's from this pool that the next generation will be constructed.

Differences in phenotype and genotype within a population are called *polymorphisms*. Some populations show a great deal of variability, while others show very little.

One method of measuring this variability is by analyzing *genotypic frequencies*, which shows which genotypes are the most or least prevalent in the population.

The genotype being analyzed is called the *locus*. Table 13-1 shows how genotypic frequencies can be analyzed for one particular genotype, where the relevant genes are labeled A and B. The possible genotypes are AA, AB, and BB, and data was collected from 7287 individuals.

From this same information you can calculate the *allelic frequency*. Since each member of the population has two alleles at this locus, the entire population consists of 14,574 (7287 × 2) alleles at this locus. To get the allelic frequency, you simply count the number of alleles of each type, then divide by the total number of alleles.

Thus, the allelic frequency for the A allele would be:

$$[(2 \times 2185) + 3623]/14{,}574 = 0.5484$$

while the allelic frequency for the B allele would be:

$$[(2 \times 1479) + 3623]/14{,}574 = 0.4516$$

The convention is to designate one of the alleles p and the other one q, so above, $p = 0.5484$ and $q = 0.5416$. ($p + q$ must equal 1, or not all alleles have

Table 13-1.   Calculating Genotypic Frequencies

Genotype	No. of Individuals	Genotypic Frequencies
AA	2185	AA=2185/7287=30.0%
AB	3623	AB=3623/7287=49.7%
BB	1479	BB=1479/7287=20.3%

been accounted for.) A population is considered to be *polymorphic* if two alleles are segregating, and the frequency of the most frequent allele is less than 0.99, as in this example.

The basic unifying concept of population genetics is the *Hardy-Weinberg Law*, named after English mathematician Godfrey Harold Hardy and Wilhelm Weinberg, a German physician, who independently formulated it in 1908. It states there will be no change in the allelic frequencies from generation to generation provided a certain number of conditions are met. A population in that condition is said to be in *equilibrium*.

The conditions that must be met to establish Hardy-Weinberg equilibrium are:

1. *The population is infinitely large and mates at random.* This avoids genetic drift, a change in genotype frequency from chance deviation. One cause of this is the *founder effect*, the difference between the genotype frequencies in the gene pool of a species as a whole and the genotype frequencies of a small, isolated population of that species. This is of much less concern in moderate or large populations. So, in practice, a population can be in Hardy-Weinberg equilibrium for one or more generations even if it isn't infinitely large—which, of course, no populations are.

2. *Natural selection isn't involved.* That is, every genotype under consideration can survive just as well as any other, and is just as efficient at producing progeny.

3. *The population is closed.* In other words, individuals are neither immigrating nor emigrating. (Again, this is rare in the real world, but if the inflow and outflow of new individuals is very small, equilibrium can be maintained for some time.)

4. *There is no mutation from one allelic state to another.* This is the case unless they cancel each other out, in which case they are allowed.

5. *Meiosis is normal.* That means chance is the only factor in the combination of genes at each mating.

Evolution arises when one of these assumptions is violated, taking a population out of equilibrium. Natural selection kicks in as some individuals, with certain genotypes, live longer (and thus have more chance to reproduce), or simply prove to be more fertile. This changes the mating distribution so that it is no longer random and, as a result, genotype frequencies begin to shift in succeeding generations.

There are three primary methods of change that drive the evolution of populations: *mutation*, *migration*, and *selection*.

MUTATION

The genetic variation that gives natural selection something to work with arises primarily through mutation.

Mutations fall into one of three classifications, based on their effect on the phenotype:

1. *Detrimental.* Most mutations are detrimental, and most detrimental mutations are eliminated from the population because individuals born with a detrimental mutation are less likely to survive or pass on their genes to offspring. However, there are examples of mutations that are detrimental in a homozygous state surviving in a population because they are beneficial in a heterozygous state. Sickle-cell anemia is one such example. Although, when homozygous, the sickle-cell mutation causes disease and often early death, individuals who are heterozygous for it are resistant to infection by the parasite that causes malaria. Thus, individuals in geographic regions where malaria is endemic are more likely to survive and reproduce than those without a copy of the gene.

2. *Neutral.* Neutral mutations, which don't provide any advantages or disadvantages to an organism, aren't affected by natural selection. Genetically, such a mutation either alters a codon in a way that doesn't change the amino acid coded for, or changes it to code for an acceptable alternative. Neutral mutations tend to be eliminated from a population through genetic drift.

3. *Beneficial.* These rare mutations convey some kind of advantage to individuals, enhancing their ability to survive or to reproduce. These are the ones that tend to become fixed in the population, replacing the nonmutated, but less fit allele.

Since more mutations are detrimental than beneficial, and most seriously detrimental mutations are quickly eliminated from the gene pool (or never even get into it), most genetic variation in a population will consist of either neutral mutations or those that might be mildly detrimental but have very little affect on fitness.

One modification of the genome that can occur is *gene duplication*. If an important gene is duplicated, mutation in the duplicated copy won't necessarily affect the individual's fitness, because he or she still has a working copy of the original gene. That allows for further mutation of the new copy of the gene, to the point where an entirely new gene may develop which has a similar function to the original but may function at a specific time in development or a unique location. This leads to *multigene families*. Hemoglobin and muscle genes in humans are organized as multigene families, as are seed storage and photosynthesis genes in plants.

MIGRATION

Migration changes gene frequencies in a population by bringing in more copies of one particular allele or by introducing a mutation that has arisen in another population of the same species. (Without this migration, isolated populations of the same species may, over time, evolve into two different species.)

In everyday usage, migration occurs as soon as an individual arrives in a new population. In genetic terms, migration only occurs when that individual successfully mates with someone in the new place, and thus introduces his or her genes into the population. This introduction of new genes through migration is called *gene flow*.

SELECTION

As already discussed, if individuals in a population have a particular genotype that makes them better able to survive and reproduce—and can pass that advantage on to their offspring—then over time that genotype will become more frequent in the population.

The Origin of Species

A *species* can be defined, in population genetics terms, as "a group of populations through which genes can flow and whose offspring have a fitness equal to the parents."

If gene flow (that is, migration) is stopped between one population of a species and all the other populations of the species, the isolated population will undergo different variations in genotype frequencies than the other populations. Over time, it can diverge so much that its members can no longer reproduce with members of the other populations. At that point, gene flow is no longer even possible, and the isolated population has become a new species. This is called *phyletic evolution* or *anagenesis*.

It's also possible for the isolated population to evolve into two distinct species that exist simultaneously. This is called *true speciation* or *cladogenesis.*

Cladogenesis has taken place when two subpopulations are no longer able to interbreed. There are three types of cladogenesis:

1. *Allopatric speciation.* This occurs by the same mechanism as phyletic evolution, but instead of an entire population being cut off by a physical barrier, it's a subpopulation that is isolated (becoming, in effect, a new, small population).
2. *Parapatric speciation.* This can occur when a subpopulation migrates into a new ecological niche not previously inhabited by the species.
3. *Sympatric speciation.* This occurs when a subpopulation develops a mutation that prevents it from mating with the original population (but not with other members of the subpopulation) and is better adapted to a particular ecological niche than the originals. For example, there's a saltmarsh species called *Spartina townsendii* that was derived from *S. alterniflora* (American saltmarsh grass) and *S. maritime* (European saltmarsh grass), but can't reproduce with either of them. It seems to be better adapted to the coastal regions of Holland than either of its parent species, and thus has established itself there.

Phylogenetic Systematics

Taxonomy is the field of science that classifies life into groups. It was pioneered by Carolus Linnaeus, who instigated the familiar *binomial nomenclature* (two-word naming) we use to identify organisms, *i.e., Homo sapiens.* Today, taxonomy has been subsumed into a larger field of science called *systematics*, which attempts to unravel the relationships among various forms of life.

Charles Darwin recognized that the taxonomy already in place by the time he wrote his famous books represented a rough approximation of evolutionary history (even though Linnaeus didn't have a clue about evolution). But it wasn't until the 1950s that a German entomologist, Willi Hennig, proposed that systematics should in fact be based on the known evolutionary history of organisms as much as possible. He called this approach *phylogenetic systematics*, and the emphasis was not on species but on *monophyletic groups*—a group plus all of its descendants—otherwise known as *clades.* This approach is now often referred to as *cladistics.*

Scientists, using clues provided by the fossil record and genetic information, use *phylogenetic trees* whose *topology* (branching pattern) represents the relationships among various organisms (see Fig. 13-1).

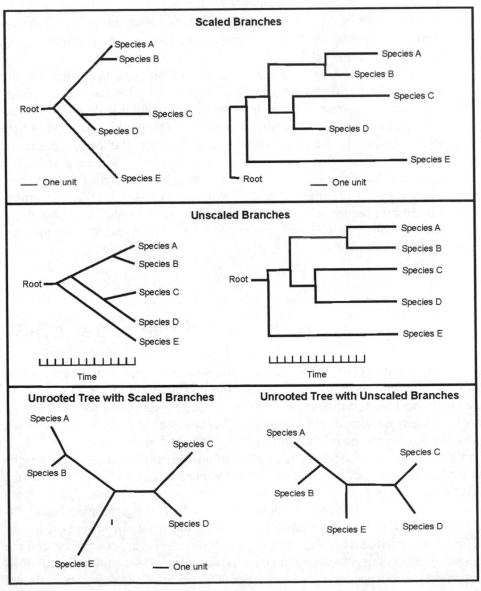

Fig. 13-1. Several different ways to draw a phylogenetic tree.

Phylogenetic trees have several parts:

- *Nodes*. These represent a taxonomic unit (species, populations or individuals, existing or ancestral).
- *Branch*. This defines the relationship between the taxonomic units in terms of descent and history.
- *Root*. This is the common ancestor of all the taxonomic units.
- *Clade*. A group of two or more taxonomic units or DNA sequences that include both their common ancestor and all of their descendants.

Only one branch can connect any two adjacent nodes. Branches can be *scaled* or *unscaled*; if the branches are scaled, the length of the branch usually indicates the number of changes that have taken place. They can also be *rooted* or *unrooted*; in rooted trees, there is a particular node that represents a common ancestor, while in unrooted trees, the relationship among taxonomic units is indicated but no common ancestor or evolutionary path is identified.

The ability to sequence DNA and protein has given rise to *molecular phylogenetics*, which uses detailed genetic information to determine the relationships among various organisms, especially those whose external form and inner structure (morphology) does not provide enough information for classification to be made. The focus on molecular phylogenetics is not on the evolution of whole organisms (although information about that evolution can be determined by molecular phylogenetics), but on the evolution of specific sequences of DNA.

The idea is that, since genomes evolve through the gradual accumulation of mutations, two genomes that are very similar may be presumed to have diverged more recently than two genomes that are very different.

The primary technique is the comparison of *homologs*, DNA sequences that have common origins but may or may not have common activity. Sequences that share some level of similarity, based on the alignment of base pairs, are said to be *homologous* (or *monophyletic*), and are presumed to have been inherited from a common ancestor, although modification over time may make it difficult to determine what that common ancestor might have been.

There are three kinds of homologs:

- *Orthologs* are homologs produced by speciation. They are genes that diverged from a common ancestor because the organism diverged, and they tend to have similar functions in whatever organisms have them.
- *Paralogs* are homologs produced by gene duplication. These are genes, derived from a common ancestor, that duplicated within an organism and then diverged. As noted earlier, duplicate genes can mutate more freely

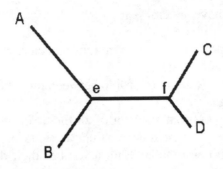

Fig. 13-2. A typical rootless phylogenetic tree based on homologs.

without lessening the organism's viability. As a result, they tend to have different functions in different organisms.

- *Xenologs* are homologs that result from the horizontal transfer of a gene between two organisms. Xenologs' functions can vary from organism to organism, but in general tend to be similar.

Fig. 13-2 shows a typical gene-based phylogenetic tree, or *gene tree* for short. There are four nodes, each representing one of the four genes, and two internal nodes representing ancestral genes. The branch lengths indicate how much evolutionary difference there is between the various genes. This tree is rootless—it says nothing about the evolutionary pathways the nodes followed.

Fig. 13-3 shows three rooted trees that could be created from the information in Fig. 13-2. Each implies a different evolutionary pathway, but there's no way to choose among them.

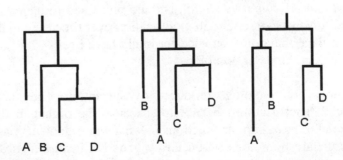

Fig. 13-3. Three different rooted phylogenetic trees, each of which could be reasonably inferred from the tree in Fig. 13-2.

For the most likely evolutionary pathway to be discerned, a gene tree needs to include at least one *outgroup*, a gene less closely related to the four homologs under study than they are to each other. Outgroups enable the root of the tree to be identified and the correct evolutionary pathway to be described (see Fig. 13-4).

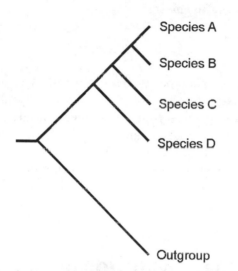

Species A

Species B

Species C

Species D

Outgroup

Fig. 13-4.   An outgroup—a gene known to have branched away from the four homologs under consideration before their common ancestor appeared—can help pin down the proper root for the phylogenetic tree in Fig.13-3.

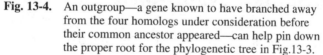

The Molecular Clock Hypothesis

Central to the practice of molecular phylogenetics is the hypothesis that point mutations—nucleotide substitutions—occur at a constant rate, so that the degree of difference between two sequences can be used to determine the date at which their ancestral sequence diverged.

However, the rate of molecular change differs among different organisms, among genes, and even among different parts of the same gene.

That means molecular clocks have to be calibrated with known fossils to determine the timing of the origin of clades and the rate of molecular change since then. For example, the molecular clock for mitochondrial DNA (mtDNA) was set by the human–chimpanzee divergence of six million years ago. Because human and chimp mtDNA differs by about 12 percent, the rate of change for human mtDNA is set at two percent per million years.

The necessity of calibrating molecular clocks means the fossil record is still vitally important to the practice of systematics. It hasn't been replaced by genetic analysis by any means.

Note that gene trees and species trees don't always match up. An internal node in a gene tree indicates the divergence of an ancestral gene into two genes with different DNA sequences from each other. An internal node in a species tree represents the divergence of an ancestral species into two species that can no longer interbreed. Since every split of a gene does not, by any means, herald the start of a new species, the two events don't always occur at the same time.

The Genetics of Human Evolution

In Chapter 10, I wrote about the "African Eve" hypothesis, based on evidence found in human mitochondrial DNA that all humans alive today can be traced back to a woman who lived in Africa about 200,000 years ago. (Note there is a fairly large margin of error in the mitochondrial DNA clock used to arrive at that estimate; some new evidence indicates Eve should be dated at 400,000 years ago, and the upper end of the possible age range is 800,000 years ago.)

Genetics, however, can and has provided information about our even more distant ancestors.

The same mitochondrial DNA used to trace our common ancient maternal ancestor first answered a longstanding evolutionary question about whether humans were more closely related to gorillas or chimpanzees. We are, it seems certain, most closely related to chimpanzees. Our most recent common ancestor with them lived five to six million years ago, whereas our most recent common ancestor with gorillas lived about eight million years ago (see Fig. 13-5).

DNA can also tell us more about human diversity. As noted in Chapter 6, humans have around three billion base pairs in their genome, and in any two individuals, 99.9 percent of the genome will be identical. Of course, 0.1 percent of three billion is still three million base pairs, and even though most of the variations make no measurable difference, we still differ from each other in thousands of ways.

But we actually have less genetic variation than our closest relatives do. Any two chimpanzees have four times as much genetic variation between them as any two humans. This probably indicates that at some time in the past, humans either underwent a severe but shortlasting population crunch that cut the population to just a few thousand, or a longerlasting, more moderate population crunch which cut the population to just a few tens or hundreds of thousands.

Interestingly, although race might seem like the most obvious way in which our genetic variety is manifested, analysis of DNA sequence data has confirmed work from the 1970s based on sequencing proteins: Fully 80 to 90 percent of variation occurs within ethnic populations, while five to 10 percent is among ethnic populations within the major racial groups, and only five to 10 percent is among the major racial groups.

Or, to put it another way, if all humans became extinct except one East African tribe, 85 percent of all human variation would still be preserved as the population rebounded.

Human genetic variation plays a large role in both disease resistance (which has helped shape our evolution through natural selection) and in the development of diseases with a genetic component.

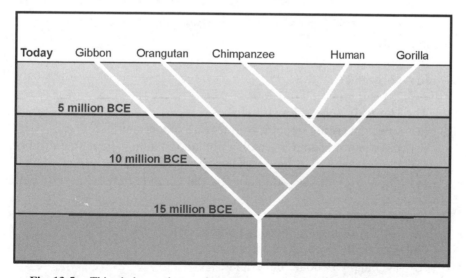

Fig. 13-5.   This phylogenetic tree shows how we are related to monkeys and apes.

In the final chapter of this book, we'll look at some of these genetic diseases, and how they—and our increasing knowledge of how to screen for them and other genetic variations—affect us now and will affect us in the future.

Quiz

1. Darwin called the driving force of evolution
   - (a) natural perfection.
   - (b) natural detection.
   - (c) natural election.
   - (d) natural selection.

2. The gene pool is
   - (a) a hypothetical mixture of all genes within a population.
   - (b) a laboratory technique for collating genes from several individual organisms.
   - (c) genes that existed in ancestral organisms but exist no longer.
   - (d) the fluid inside a cell nucleus in which DNA is located.

3. A population that has one or more alleles of a particular gene is said to be
   - (a) polyannish.
   - (b) polyester.
   - (c) polymorphic.
   - (d) polygenic.

4. The Hardy-Weinberg Law describes
   - (a) rules governing genetic engineering in Germany.
   - (b) the conditions under which a population attains genetic equilibrium.
   - (c) the mechanism of natural selection in insects.
   - (d) the maximum size attainable by the genome of a eukaryotic cell.

5. Neutral mutations
   - (a) usually disappear through genetic drift.
   - (b) eventually become detrimental.
   - (c) eventually become beneficial.
   - (d) never occur.

6. In genetic terms, migration doesn't take place until an immigrating organism
   (a) buys a house.
   (b) gets a driver's license.
   (c) successfully mates.
   (d) mutates.

7. For a population to become a new species, it must be
   (a) isolated.
   (b) irradiated.
   (c) invaded.
   (d) irritated.

8. The field of science that examines the evolutionary relationship between organisms is called
   (a) evolutionism.
   (b) relational genetics.
   (c) systematics.
   (d) nomenclature.

9. DNA sequences in different organisms that share a common ancestry, but not necessarily a common function, are called
   (a) homologs.
   (b) homogenes.
   (c) homoclades.
   (d) homobases.

10. Genetics reveals our closest animal relatives are
   (a) monkeys.
   (b) gorillas.
   (c) chimpanzees.
   (d) lemurs.

Humans—How Genetics Affects Us

The title of this chapter is, admittedly, misleading. Genetics don't *affect* us; it would be more accurate to say that "Genetics *'R'* Us." But there's no question that our enhanced and increasing *knowledge* of genetics does affect us, and will affect us in the future, perhaps in ways we have yet to imagine.

The importance of genetics to humans begins at the very beginning: with conception.

The Genetics of Sex

Although the focus of this chapter is primarily on humans, and despite what we all think in adolescence, we did not invent sex. Nor is our familiar method of combining genes the only one invented by nature. There's a version of the protozoan, *Paramecium bursaria*, for example, that has eight sexes, each incapable of mating with its own sex, but able to exchange genetic information with any of the other seven.

However, in most higher animals, the sexes are limited to two (although sometimes both sexes are present in the same organism: An animal with both male and female reproductive organs is called a *hermaphrodite*, while a plant with both is called *monoecious*).

Genetically, these details don't matter. The function of sex is to create genetic variability through the exchange of genetic information in order to insure a population can adapt to changing environmental conditions.

The first genetic step toward this eventual exchange is to determine which sex an organism is (in those species in which the sexes are different organisms, such as *Homo sapiens*).

In most mammals there are two different, or *heteromorphic*, sex chromosomes, the X and the Y chromosomes. The presence of the Y chromosome makes an organism male. So females have an XX pair of chromosomes, and their gametes always contain an X chromosome, while males have an XY pair, and their gametes may contain either an X or a Y chromosome. Females, in other words, are *homogametic*, while males are *heterogametic*.

The male gamete therefore determines the sex of the embryo that forms from its successful fertilization of an egg. (There are some insects, birds, and fishes in which it is the female who is heterogametic and the males are homogametic. To avoid confusion, the relevant chromosomes in those species are sometimes labeled Z and W instead of X and Y.)

At least one specific gene that helps determine sex has been identified. *SRY* (for sex-determining region Y), located on the short arm of the Y chromosome, encodes a gene product called the *testis-determining factor (TDF)*. In conjunction with several other genes, it encodes a DNA-binding protein that, in turn, apparently controls the expression of one or more other genes that are involved in testicular development and fertility.

Because of the way sex is determined in humans, and because of the very different sizes of the two chromosomes involved (X codes for 2000 to 3000 genes, while Y codes for only a few dozen), there are many traits whose genes appear only on the X chromosome and have no corresponding genes on the Y chromosome. There are also a few whose genes appear only on the Y chromosome and have no corresponding genes on the X chromosome, but these are rare.

Sex-linked traits may be either dominant or recessive. A trait governed by a sex-linked recessive gene on the X chromosome will typically have three characteristics which are quite different from those of traits governed by genes found on the *autosomes* (nonsex chromosomes):

1. It is usually found more frequently in the male than in the female.
2. It fails to appear in females unless it also appears in the paternal parent.

3. It seldom appears in both father and son, and then only if the mother is heterozygous.

Perhaps the best known example of this kind of sex-linked trait is hemophilia, a disease characterized by a failure of the blood to clot properly (see Table 14-1 for others).

The X chromosome carries the recessive genes that control the production of two blood-clotting substances called clotting Factors VIII and IX. If a girl inherits an X chromosome with missing or poor copies of these genes, she will usually still have copies of the genes on her other X chromosome, and so won't have hemophilia.

However, if a boy inherits an X chromosome with missing or poor clotting-factor genes, he has no backup genes—they don't exist on the Y chromosome. As a result, his blood is missing those clotting factors, and he has hemophilia.

Although it is extremely rare, it is possible for a girl to be born with hemophilia if her mother carries one copy of the faulty X chromosome and her father has hemophilia.

A trait governed by a sex-linked dominant gene on the X chromosome usually shows these characteristics:

1. It is usually found more frequently in the female than in the male.
2. It is found in all female offspring of a male that shows the trait.
3. It isn't transmitted to any son from a mother that didn't exhibit the trait herself.

Table 14-1.   Examples of Hereditary Sex-linked Conditions Other than Hemophilia

Disease	Symptoms
Duchenne's muscular dystrophy	Progressive muscle weakness followed by death
Lesch-Nyhan syndrome	Cerebral palsy, self-mutilation
Anhidrotic ectodermal dysplasia	Inability to sweat, absent teeth, sparse hair
Agammaglobulinemia	Absence of immunoglobulins, frequent severe bacterial infections
Fragile-X syndrome	Mental retardation, characteristic facial structure, large testes

The X and Y chromosomes couldn't pair during meiosis if they didn't have at least some matching DNA. This section is called the *human pseudoautosomal region*. One gene known to be located here is *MIC2*, which codes for a glycoprotein used in the membranes of certain body cells. The genes in this region behave just like the genes on the nonsex chromosomes and follow the usual Mendelian pattern of inheritance.

Only about a dozen genes reside in the nonrecombining part of the Y chromosome, even though it makes up about 95 percent of the chromosome. Such traits are expressed only in males, and always transmitted from father to son. *SRY*, mentioned earlier, is one of these; you might say (in fact, I think I will) that maleness is itself the ultimate sex-linked trait.

In addition to sex-linked traits, there are also a number of *sex-influenced* traits. The genes that govern these traits may reside on the autosomes or in the pseudoautosomal region of the sex chromosomes. Due to the influence of the sex hormones, the normal expression one would expect from dominant or recessive alleles is reversed in sex-influenced traits. These traits are therefore usually found in higher animals with well-developed hormonal systems— like humans.

Probably the best known human example of this is pattern baldness, which acts like a dominant trait in men, but like a recessive one in women (for which women are thankful). Table 14-2 illustrates this. When a man has one gene for baldness and one for nonbaldness, he goes bald, but when a woman has the same combination, she doesn't.

In addition to sex-influenced traits, there are also *sex-limited* traits. These are traits that are only expressed in one of the sexes, even though the genes are present in both, either because of hormonal control or anatomical differences. For example, men carry genes for milk production which they pass on to their daughters, but they don't make milk themselves.

Table 14-2. Pattern Baldness, a Sex-influenced Trait

	Phenotypes	
Genotypes	**Men**	**Women**
Two copies of bald gene	Bald	Bald
One bald gene, one nonbald gene	Bald	Non-bald
Two nonbald genes	Non-bald	Non-bald

Genetic Diseases

In 1902, a British doctor named Archibald Garrod showed that a rare disease called *alkaptonuria*, which turns the urine of those who have it black, was inherited according to Mendel's rules. It was the first human disease to be shown to be inherited in a Mendelian pattern—in fact, it was the first human trait of any kind shown to be inherited in a Mendelian pattern. Garrod called it an *inborn error of metabolism*. Today we'd call it a *genetic disease*.

Genetic diseases are those that arise from problems in the genome of the organism. These may be inherited or they may be the result of a mutation in the gametes. In the latter case, if the individual survives and reproduces, the problem can be inherited by future generations.

Among the most serious genetic disorders are those that arise because of a mistake during cell division early in development (or even before conception), causing an error in the chromosome number. These chromosomal problems include:

- *Trisomy*. In this condition, there are three copies of a particular chromosome instead of the normal two. The most familiar trisomy is trisomy 21, also known as Down syndrome.
- *Monosomy*. In monosomy, one member of a chromosome pair is missing. An example is Turner syndrome, a disorder affecting girls born with only one X chromosome.
- *Deletions*. A small part of a chromosome is missing; in *microdeletions*, the missing part amounts to only a single gene or just a few genes.
- *Translocations*. Bits of chromosomes are shifted from one chromosome to another.
- *Inversions*. Small parts of the chromosome's DNA sequence are snipped out and flipped over.

The effects of these problems vary widely, from negligible to fatal, depending on which chromosomes or sections of chromosomes are involved.

Mutations involving a single gene are also responsible for a wide variety of disorders. Among the diseases caused by problems with individual genes are cystic fibrosis, sickle-cell anemia (mentioned several times already), Tay-Sachs disease, and achondroplasia (a form of dwarfism).

Cancer, as was discussed in detail in Chapter 8, is also a genetic disease—both a disease caused by alterations in a cell's genes and a disease whose development may be influenced by mutations in the genome. Other diseases may not be caused solely by genomic mutations, but mutations may play a role in their

Table 14-3. Some Relatively Common Genetic Diseases

Disease	Approximate Incidence among Live Births
Cystic fibrosis	1/1600 Caucasians
Duchenne muscular dystrophy	1/3000 boys (X-linked)
Gaucher disease	1/2500 Ashkenazi Jews; 1/75,000 others
Tay-Sachs disease	1/3500 Ashkenazi Jews; 1/35,000 others
Essential pentosuria (a benign condition)	1/2000 Ashkenazi Jews; 1/50,000 others
Classic hemophilia (defective clotting factor VIII)	1/10,000 boys (X-linked)
Phenylketonuria	1/5,000 Celtic Irish; 1/15,000 others
Cystinuria	1/15,000
Metachromatic leukodystrophy	1/40,000
Galactosemia	1/40,000
Sickle-cell anemia	1/400 U.S. blacks (much higher in some West African countries)
β Thalassemia	1/400 among some Mediterranean populations

development. Parkinson's disease, for instance, may be linked to a gene on chromosome 4; multiple sclerosis may be linked to a gene on chromosome 6; and Alzheimer's may be linked to a gene on chromosome 19.

Table 14-3 lists some other genetic disorders. There are hundreds more. Although each individual disease is rare, there are so many of them that they affect millions of people in the United States alone.

RFLPs

In order to test for a genetic defect, you have to know what gene causes the problem and where that gene is located. A powerful tool both for locating disease genes and screening individuals to see if they carry such a gene is analysis of *restriction fragment length polymorphisms*, or *RFLPs* (pronounced "riflips.")

Recall from Chapter 12 that enzymes called restriction endonucleases cleave DNA molecules at specific sequences called restriction sites. Two DNA molecules will only be broken into fragments of identical length if they are, in fact, identical. If one has a restriction site in one location and the other doesn't, the resulting fragments will be of a different length; if one molecule has more base pairs between recognition sites than the other, the resulting fragments will be of different lengths, and so on.

In practice, this means that applying a restriction endonuclease to the same section of the genomes of two different individuals of the same species will most often produce fragments of different lengths, depending on the details of each individual's genome. It also means that if you're comparing homologous regions of both chromosomes of a pair, you'll get fragments of different length if one chromosome has a deletion or insertion somewhere between recognition sites, a mutation that dispenses with a recognition site altogether, or a different allele of a gene located between recognition sites.

RFLPs are inherited according to Mendelian rules, and they have proven invaluable in pinning down the location of various disease genes—such as the one that produces Huntington's disease. Researchers look in populations with a high incidence of the disease for RFLPs that are consistently inherited by those who have the disease and not present in those that do not. That's a good indication that the gene in question is located somewhere near that RFLP.

For example, RFLPs enabled the precise location of the cystic fibrosis gene to be mapped (see Fig. 14-1). Once a disease gene is located it can be cloned and studied with a goal of better understanding the defect involved in the disease and developing new therapies.

Analysis of RFLPs can be carried out using even the very small amounts of DNA obtainable from a sample of white blood cells or fetal cells, which has also made them valuable as a diagnostic tool. Knowing a particular RFLP is associated with an inherited disease gene makes it possible for doctors to diagnose conditions more quickly and accurately.

In rare but extremely useful cases, a mutant gene directly alters an RFLP. In the case of sickle-cell disease, for instance, there is a restriction site in the normal gene that does not appear in the abnormal gene (see Fig. 14-2). As a result, the normal gene creates two RFLPs, while the abnormal one creates only one.

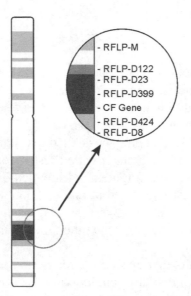

- RFLP-M
- RFLP-D122
- RFLP-D23
- RFLP-D399
- CF Gene
- RFLP-D424
- RFLP-D8

Fig. 14-1. RFLPs enabled the precise location of the cystic fibrosis gene to be determined.

The whole process of using RFLPs to map, diagnose and identify faulty genes is similar to the following, which was the procedure used to uncover the mechanism behind Duchenne's muscular dystrophy:

1. Large families known to carry the disease are screened for known and mapped RFLPs.
2. The results are searched for statistically significant links between inherited RFLPs and the disease.
3. Success means the gene can be mapped to a specific region of a specific chromosome. If several known RFLPs surround the gene and are close enough, the gene can be cloned.
4. If the normal gene can be cloned, a probe can be created from it and used in a library constructed from normal muscle cells.
5. From the library, the normal gene and any other necessary sequences for its expression can be cloned.
6. The resulting DNA can be inserted into *E. coli* via a vector.
7. *E. coli* synthesizes the resulting protein. This is the protein the normal gene produces.
8. Tests can then determine whether those suffering with the disease have this protein and, if so, where it is and isn't expressed. In the case of Duchenne's muscular dystrophy this process showed that the disease results from the

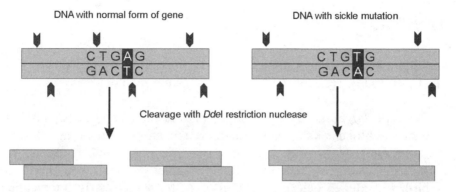

DNA with normal form of gene   DNA with sickle mutation

Cleavage with *Dde*I restriction nuclease

Fig. 14-2. An RFLP that appears in a normal hemoglobin gene does not appear in the faulty gene that causes sickle-cell disease, allowing the disease to be diagnosed by this method.

body's inability to synthesize a protein called *dystrophin*, due to a mutated dystrophin gene.

This same procedure is being followed in research efforts aimed at other poorly understood genetic diseases.

Forensic Genetics

A tangentially related use of restriction nucleases is for *forensic genetics*, used to establish the identity of DNA left behind at a crime scene or taken from a suspect or victim.

DNA fingerprinting relies on the fact that different human genomes have different numbers of tandem repetitive DNA sequences. Each of these is called a *minisatellite* or (much more wordily) a *variable number of tandem repeats locus* (VNTR locus).

The number, pattern, and length of these repeating sequences are unique for each individual. But no matter how long the sequence is, it contains a common core sequence—usually less than 20 base pairs in length—that can be recognized by an appropriate probe.

DNA from a sample taken from the crime scene or the suspect's or victim's cells is cloven by one or more restriction endonucleases, then separated on a gel, denatured to single strands, transferred to a nitrocellulose filter, exposed to a radioactively labeled probe, and autoradiographed. The number of bands that show up on the autoradiograph are unique for each individual.

Genetic Screening

Our ability to examine individuals' genomes for potential problems is increasing all the time. Today, areas of focus in genetic testing include:

- *Prenatal diagnosis.* By examining amniotic fluid, fetal cells, and fetal or maternal blood cells, doctors can discern whether a fetus is at risk for various identifiable genetic diseases or traits, such as Down syndrome.
- *Newborn screening.* Blood or tissue samples can be taken shortly after birth in order to detect genetic diseases for which early intervention is necessary to prevent serious health problems or death. For example, since the 1970s, newborn African-American infants have been tested for sickle-cell anemia.
- *Carrier screening.* This identifies individuals with a gene or chromosomal abnormality that may cause problems for offspring or the person screened. Tests have been developed for cystic fibrosis, Duchenne muscular dystrophy, hemophilia, Huntington's disease, and others.
- *Susceptibility testing.* This is used to identify workers who may be susceptible to toxic substances found in their workplace, and hence at risk of future disabilities.

While the promise of genetic screening to lessen human suffering is obvious, there are many concerns about how the information might be used—especially if it might move beyond the patient/doctor realm and into the realm of business, where a person might be refused insurance, for example, because of a genetic tendency toward heart disease at a fairly early age.

As a poll taken by the Genetics and Public Police Center at Johns Hopkins University in 2002 revealed (see Table 14-4), these concerns over possible *genetic discrimination* are widespread in the public.

Table 14-4. To Which of These Would You Be Willing to Show Your Genetic Test Results?

	Yes	No
Husband, wife or partner	68%	32%
Immediate family	53%	47%
Insurance companies	27%	73%
Employer	12%	88%

Some are calling for stiff laws governing the release of genetic testing information. Others warn that with research discovering that almost every disease has a genetic component—findings that are likely to accelerate with the influx of new information generated in part by the completion of the Human Genome Project and in part simply by the continuing advances in understanding of every aspect of genetics—any law could have unintended consequences. For example, health insurers may be denied the basic information they need to assess the risk pool and set adequate rates.

The more we learn about genetics, it seems, the more these kinds of ethical questions surface. We may see, in the next few decades, human cloning, designer babies, widespread genetic screening, genetically modified humans, effective gene therapy, outstanding new cancer treatments, and new vaccines, among many other developments. The possibilities—for both human health and ethical pitfalls—are endless.

When Gregor Mendel first planted peas in the monastery garden in Brunn in the mid-19th century, little did he know that at the start of the 21st century, we'd still be harvesting his bounty.

Quiz

1. Animals with two different sex chromosomes are called
   (a) heteromorphic.
   (b) heliotropic.
   (c) heterogenic.
   (d) homophobic.

2. In mammals, most sex-linked genes are located on
   (a) the W chromosome.
   (b) the X chromosome.
   (c) the Y chromosome.
   (d) the Z chromosome.

3. Autosomes are
   (a) chromosomes that can replicate outside a cell.
   (b) chromosomes that contain the genes for being a good driver.
   (c) chromosomes that aren't sex chromosomes.
   (d) sex chromosomes.

4. The chromosomal region shared by both the X and Y chromosomes in humans is called
   (a) the generic human region.
   (b) the human asexual region.
   (c) the chromosomal agreement region.
   (d) the human pseudoautosomal region.

5. Today, an "inborn error of metabolism" is usually called
   (a) the sex drive.
   (b) a genetic disease.
   (c) the common cold.
   (d) indigestion.

6. Someone with three copies of a chromosome instead of the usual two is said to suffer from
   (a) trisomy.
   (b) trigenics.
   (c) chromotrinity.
   (d) somecubing.

7. RFLP stands for
   (a) radically foreign lipid proteins.
   (b) restructured folic-lysing protocols.
   (c) really funny long pronouns.
   (d) restriction fragment length polymorphisms.

8. Which of the following is true?
   (a) RFLPs are inherited according to Mendelian rules.
   (b) RFLPs are always exactly the same length.
   (c) RFLPs are useless for locating genes on chromosomes.
   (d) RFLPs never vary from individual to individual.

9. VNTR locus is a term used in
   (a) microscopy.
   (b) cell culturing.
   (c) DNA fingerprinting.
   (d) football.

10. Americans are least willing to share the results of their genetic testing with
    (a) husband, wife, or partner.
    (b) immediate family.
    (c) insurance companies.
    (d) employer.

Final Exam

Do not refer to the text while taking this quiz; 75 percent correct is a good score. Answers are in the following section. You might want to have a friend check your score for you the first time, so you don't just memorize the answers and can take the test again.

1. Whose work with garden peas is the basis of modern genetics?
   (a) Gregor Mendel
   (b) Robert Hook
   (c) Isaac Newton
   (d) James Watson

2. The principle of segregation states that
   (a) species cannot interbreed.
   (b) plant seeds must be planted a specific distance from the parent plant before they will sprout.
   (c) each parent gives only one of its hereditary "factors" to its offspring.
   (d) the source of heredity is ultimately unknowable.

3. The principle of independent assortment states that
   (a) one trait in an organism has no bearing on another.
   (b) no general genetic principles can be inferred from the study of a single organism.
   (c) two different species of plants grown side by side may exchange genetic information without breeding.
   (d) genetic factors are spread equally among all cells of an organism.

4. If an individual organism is heterozygous, it has
   (a) one form of a particular gene.
   (b) two different forms of a particular gene.
   (c) characteristics of both sexes.
   (d) a fatal genetic disease.

5. One of the possible forms of a gene is called
   (a) a zygote.
   (b) a chromosome.
   (c) an allele.
   (d) a factor.

6. Eukaryotic cells are characterized by
   (a) photosynthesis.
   (b) the ability to move on their own.
   (c) genetic material scattered throughout the cell body.
   (d) a well-defined nucleus and other membrane-bound structures.

7. Prokaryotic cells are characterized by
   (a) an ATP metabolism.
   (b) lack of genetic material.
   (c) a single outer membrane and no defined nucleus or other internal structures.
   (d) immobility.

8. Organelles are
   (a) filaments that form between chromosomes.
   (b) small structures within the cell that perform various tasks.
   (c) light-sensitive spots found in some prokaryotes.
   (d) deep creases in the cell's membrane.

9. The duplication and division of a body cell's chromosomes is called
   (a) mitosis.
   (b) meiosis.
   (c) halitosis.
   (d) spindling.

10. Meiosis is
    (a) the duplication and division of a body cell's chromosomes.
    (b) the cell-division process that produces gametes (sperm and eggs)
    (c) the construction of a new cell membrane just before two daughter cells split.
    (d) the programmed death of a cell after a certain number of divisions.

11. A haploid cell contains
    (a) only half the full complement of chromosomes.
    (b) the usual complement of chromosomes.
    (c) twice the full complement of chromosomes.
    (d) damaged chromosomes.

12. Oswald Avery's work confirmed that
    (a) DNA is the organizing principle of heredity.
    (b) sex in humans is determined by the X and Y chromosomes.
    (c) fruit flies and humans share the same genetic material.
    (d) some diseases are caused by substances smaller than bacteria.

13. The backbones of the DNA double helix are made of
    (a) amino acids.
    (b) RNA.
    (c) sugars and phosphates.
    (d) lipids.

14. A nucleotide is
    (a) one piece of DNA backbone with a base attached.
    (b) a strand of DNA labeled with a radioactive tag.
    (c) a short strand of RNA used to transfer information between DNA strands.
    (d) a special type of chromosome.

15. A replication fork is
    (a) the split caused by binary fission in cells.
    (b) a laboratory instrument used to stimulate the replication of DNA.
    (c) the point at which replication is taking place in a DNA strand.
    (d) a genetic defect found in some gametes.

16. Enzymes are
    (a) proteins that act as catalysts, allowing biochemical reactions to take place without themselves being changed.
    (b) small subsections of chromosomes that can be stained and tracked throughout the process of cell division.
    (c) sections of DNA that do not seem to serve any hereditary purpose.
    (d) toxins that break down cell structures, eventually killing the cell.

17. Sex cells have
    (a) the same number of chromosomes as body cells.
    (b) half as many chromosomes as body cells.
    (c) twice as many chromosomes as body cells.
    (d) one quarter as many chromosomes as body cells.

18. *Drosophila melanogaster* is a type of
    (a) virus.
    (b) bacteria.
    (c) slime mold.
    (d) fruit fly.

19. A trait that only occurs in one gender is called a
    (a) sex-limited trait.
    (b) nonsexual trait.
    (c) generic trait.
    (d) recessive trait.

20. How many individual chromosomes do human body cells have?
    (a) 23
    (b) 46
    (c) 12
    (d) 69

21. The centromere
    (a) "unzips" DNA molecules.
    (b) appears between two daughter cells during binary fission.
    (c) determines the location of the nucleus in a cell.
    (d) divides chromosomes into two arms of varying lengths.

22. Genes that are found on the same chromosome are said to be
    (a) linked.
    (b) bound.
    (c) intertwined.
    (d) bipolar.

23. Mapping chromosomes reveals
    (a) genetic abnormalities.
    (b) their location in the cell nucleus.
    (c) what kind of cell is being examined.
    (d) where specific genes are located.

24. DNA can be read as a code for producing a chain of
    (a) cells.
    (b) sugars.
    (c) amino acids.
    (d) salts.

25. A codon is
    (a) one three-letter "word" in the genetic code.
    (b) a protein cap on the end of a chromosome.
    (c) the enzyme that makes DNA replication possible.
    (d) a special kind of RNA that provides energy for cell fission.

26. Transcription is
    (a) the exchange of genetic information between the members of a chromosome pair.
    (b) another word for binary fission.
    (c) the entering of DNA sequences into a computer for analysis.
    (d) the process of copying the genetic information from DNA to mRNA.

27. Translation is
    (a) another word for binary fission.
    (b) the conversion of cell organelles into DNA.
    (c) the process of building chains of amino acids from the genetic information copied into mRNA
    (d) the process of rearranging a cell's DNA to change the cell's function.

28. Ribosomes are
    (a) tiny structures in the cell cytoplasm that provide energy for all cell functions.
    (b) tiny structures in the cell cytoplasm that control the process of building amino-acid chains.
    (c) tiny structures in the cell nucleus that have their own DNA separate from the nuclear DNA.
    (d) tiny structures in the cell membrane that control the flow of proteins through the cell wall.

29. The physical expression of genetic information in an organism is called its
    (a) phenotype.
    (b) genotype.
    (c) trait indicator.
    (d) protein display.

30. Polygenic traits are those determined by
    (a) non-nuclear DNA.
    (b) sex.
    (c) more than one gene.
    (d) only one gene.

31. Homeotic genes
    (a) are located in the mitochondria.
    (b) are much smaller than ordinary genes.
    (c) determine sexual orientation.
    (d) initiate or block the action of other genes.

32. The most commonly used method of sequencing DNA was developed by
    (a) Barbara McClintock.
    (b) Francis Crick.
    (c) Oswald Avery.
    (d) Fred Sanger.

33. The human genome contains approximately how many base pairs?
    (a) three thousand
    (b) three million
    (c) three billion
    (d) three trillion

34. What was J. Craig Venter's approach to genome sequencing dubbed?
    (a) the shotgun approach
    (b) the machine-gun approach
    (c) the silver-bullet approach
    (d) the nuclear option

35. About how many genes are currently thought to be in the human genome?
    (a) 2500
    (b) 25,000
    (c) 250,000
    (d) 2,500,000

36. My genome probably differs from your genome about
    (a) once in every 1200 bases.
    (b) once in every 12,000 bases.
    (c) once in every 120,000 bases.
    (d) once in every 1,200,000 bases.

37. A genetic variation that shows up in one percent or more of the population is called a
    (a) genetic migration.
    (b) evolutionary marker.
    (c) major mutation.
    (d) polymorphism.

38. A mutagen is a substance that
    (a) prevents mutations.
    (b) induces mutations.
    (c) tags mutations for analysis.
    (d) deletes mutations from DNA samples.

39. A point mutation
    (a) changes a single base pair.
    (b) deletes at least 10 base pairs.
    (c) inverts a section of a chromosome.
    (d) is always fatal.

40. A frameshift mutation
    (a) deletes an entire chromosome.
    (b) replaces one codon with a different one.
    (c) can move from place to place in the genome.
    (d) removes a base pair, offsetting the reading frame of the genetic sequence by one letter.

41. Transposons
    (a) delete entire chromosomes.
    (b) replace one codon with a different one.
    (c) can move from place to place in the genome.
    (d) remove a base pair, offsetting the reading frame of the genetic sequence by one letter.

42. A mutation that cancels out or works around the change produced by a previous mutation is called a
    (a) detrimental mutation.
    (b) neutral mutation.
    (c) benevolent mutation.
    (d) suppressor mutation.

43. The versions of genes that most often occur naturally are called
    (a) natural genes.
    (b) wild-type genes.
    (c) HF (high-frequency) genes.
    (d) LM (low-mutation) genes.

44. Cancer cells have the ability to divide indefinitely. They have become
    (a) immortalized.
    (b) desensitized.
    (c) transduced.
    (d) indestructible.

45. When cancer cells gain the ability to move independently and invade other tissues, they are said to have
    (a) evolved.
    (b) metastasized.
    (c) metamorphed.
    (d) mobilized.

46. A neoplasm is
    (a) a population of cells growing out of control.
    (b) a cell organelle that begins the process of turning a cell cancerous.
    (c) the nucleus of a cancer cell.
    (d) pain caused by cancer cells putting pressure on neurons.

47. Non-cancer-causing genes that, when altered, can cause cancer are called
    (a) prenarcogenes.
    (b) neogenes.
    (c) protooncogenes.
    (d) noncosomes.

48. Retinoblastoma is a cancer of
    (a) the eye.
    (b) the colon.
    (c) the bladder.
    (d) the lymph system.

49. Carcinogens are
    (a) genes that can cause cancer.
    (b) cancers that cause additional mutations.
    (c) mutagens that can turn cells cancerous.
    (d) toxic waste products produced by cancerous cells.

50. "A single ring-shaped DNA molecule" is a description of
    (a) the only genetic material in a cancer cell.
    (b) the bacterial chromosome.
    (c) the viral chromosome.
    (d) a transposon.

51. Without DNA gyrase, bacterial chromosomes would
    (a) become wound so tight they couldn't reproduce.
    (b) fracture.
    (c) mutate.
    (d) form chromosomes like those in eukaryotic cells.

52. Competent bacteria can
    (a) cause disease.
    (b) remain dormant for thousands of years.
    (c) incorporate naked DNA.
    (d) survive antibiotics.

53. In conjugation, two bacteria
    (a) kill each other.
    (b) exchange genetic information.
    (c) join into one giant cell.
    (d) join forces to fight viruses.

54. Plasmids are
    (a) small DNA molecules outside of the bacterial chromosome that can
        replicate on their own.
    (b) benign viruses that invade bacteria but do not harm them.
    (c) proteins used in the construction of bacterial cell membranes.
    (d) virulent viruses that quickly kill most bacterial cells.

55. Co-repressors
    (a) trigger the shutdown of gene translation.
    (b) trigger the shutdown of gene transcription.
    (c) trigger the shutdown of cell replication.
    (d) trigger the shutdown of the immune system.

56. The endosymbiont theory suggests
    (a) that organelles have evolved from bacteria.
    (b) that Eubacteria evolved from organelles in Archaea.
    (c) that organelles are bacterial parasites.
    (d) that viruses and organelles can crossbreed.

57. Chloroplasts are plant organelles that contain
    (a) chloroform.
    (b) chlorophyll.
    (c) chlorine.
    (d) tumor-causing agents.

58. The maternal parent of an organism overwhelmingly contributes
    (a) its mitochondrial DNA.
    (b) its chloroplast DNA.
    (c) its mitochondrial DNA and chloroplast DNA.
    (d) neither mitochondrial DNA nor chloroplast DNA.

59. Mitochondrial DNA indicates a female who lived perhaps 200,000 years
    ago in Africa was
    (a) the direct maternal ancestor of everyone now living.
    (b) the last common ancestor of chimpanzees and humans.
    (c) the first member of the species *Homo sapiens*.
    (d) the first human being to use language.

60. Tobacco mosaic disease was the first eukaryotic disease recognized to be caused by
    (a) bacteria.
    (b) a virus.
    (c) genetic abnormalities.
    (d) radiation.

61. The protein coat of a virus is called
    (a) an envelope.
    (b) a protomembrane.
    (c) a viral shell.
    (d) a capsid.

62. Bacteriophages are viruses that
    (a) attack bacteria.
    (b) protect bacteria from antibiotics.
    (c) cause illnesses indistinguishable from bacterial infections.
    (d) cause cancer.

63 The bursting open of the host cell is the final stage of the
    (a) bacteriomorph cycle.
    (b) lytic cycle.
    (c) lysogenic cycle.
    (d) binary fission cycle.

64. A lipid membrane picked up by a virus from its host cell is called
    (a) a capsid.
    (b) a capsomere.
    (c) an envelope.
    (d) a gel capsule.

65. Nucleases are enzymes that can
    (a) break the bonds of DNA backbones.
    (b) break the bonds of DNA base pairs.
    (c) cause cells to burst.
    (d) kill viruses.

66. A genetically identical copy of another organism is called
    (a) a genomorph.
    (b) a clone.
    (c) a clown or sport.
    (d) a monogene.

67. PCR, a technique for making many copies of a DNA sequence quickly, stands for
    (a) Polymerase Chain Reaction.
    (b) Polygenic Copy Replication.
    (c) Perfect Cell Repeating.
    (d) Pre-Chromosomal Reaction.

68. A transgenic organism is
    (a) an organism with an unusual number of transposons in its genome.
    (b) an organism that replicates without sex.
    (c) a genetically engineered organism that can pass its new genes on to its offspring.
    (d) an organism with a lot of genetic variation within populations.

69. Higher animals are cloned via
    (a) gametic chromosomal transfer.
    (b) somatic nuclear transfer.
    (c) atomic DNA division.
    (d) haploid polymerase extension.

70. A hypothetical mixture of all genes within a population is called
    (a) the total chromosomal accumulation (TCA).
    (b) the DNA agglomeration (DNA-AGG).
    (c) the gene pool.
    (d) the reproductive resource.

71. A population can only be in genetic equilibrium under conditions laid out in
    (a) the Hardy-Weinberg Law.
    (b) the Crick-Watson Law.
    (c) the Huxley-Darwin Law.
    (d) the McLintock-Morgan Maxim.

72. Most mutations are
    (a) beneficial.
    (b) neutral.
    (c) detrimental.
    (d) fatal.

73. Genetically, the influx of new individuals into a population doesn't count as migration until those individuals
    (a) mate and produce offspring with individuals of the extant population.
    (b) build homes.
    (c) interact with other species.
    (d) are physically cut off from their previous population.

74. Systematics
    (a) examines the physical relationship between chromosomes and genes.
    (b) examines the sociological relationship between males and females.
    (c) examines the evolutionary relationship between organisms.
    (d) examines the symbiotic relationship between bacteria and animals.

75. Homologs are
    (a) DNA sequences in different organisms that share a common function, but not necessarily a common ancestry.
    (b) DNA sequences in different organisms that share a common ancestry, but not necessarily a common function.
    (c) DNA sequences in a single organism that share a common function, but have a different series of base pairs.
    (d) DNA sequences in a single organism that have the same series of base pairs, but entirely different functions.

76. Heteromorphic animals have
    (a) no sex chromosomes (sex is determined by environmental factors).
    (b) two identical sex chromosomes (sex is determined by chance).
    (c) two different sex chromosomes.
    (d) multiple sex chromosomes and multiple sexes.

77. Non-sex chromosomes are called
    (a) monosomes.
    (b) archosomes.
    (c) autosomes.
    (d) phenosomes.

78. Down syndrome is an example of a chromosomal abnormality called a
    (a) trisomy.
    (b) monosomy.
    (c) deletion.
    (d) inversion.

79. Restriction fragment length polymorphisms (RFLPs) are used for
    (a) treating cancer.
    (b) genetic testing.
    (c) enhancing fertility.
    (d) developing vaccines.

80. Americans are *most* willing to share the results of their genetic testing with their
    (a) husband, wife, or partner.
    (b) immediate family.
    (c) insurance companies.
    (d) employer.

Answers to Quiz and Final Exam Questions

CHAPTER 1

1. b   2. c   3. d   4. b   5. c
6. a   7. d   8. b   9. b   10. c

CHAPTER 2

1. c   2. d   3. b   4. d   5. c
6. a   7. c   8. b   9. d   10. a

CHAPTER 3

1. c   2. b   3. c   4. c   5. a
6. b   7. a   8. c   9. d   10. b

CHAPTER 4

1. d   2. b   3. c   4. d   5. b
6. a   7. d   8. a   9. d   10. a

CHAPTER 5

1. b   2. c   3. a   4. d   5. c
6. a   7. c   8. a   9. d   10. b

CHAPTER 6

1. c   2. a   3. c   4. d   5. a
6. b   7. a   8. c   9. a   10. c

CHAPTER 7

1. d   2. c   3. a   4. b   5. c
6. b   7. a   8. c   9. a   10. b

CHAPTER 8

1. a   2. c   3. b   4. d   5. c
6. c   7. a   8. d   9. b   10. c

CHAPTER 9

1. a	2. c	3. b	4. d	5. b
6. c	7. b	8. a	9. d	10. b

CHAPTER 10

1. c	2. d	3. b	4. b	5. a
6. b	7. a	8. c	9. d	10. a

CHAPTER 11

1. c	2. b	3. a	4. c	5. b
6. d	7. c	8. a	9. c	10. a

CHAPTER 12

1. a	2. b	3. d	4. a	5. c
6. a	7. b	8. c	9. d	10. b

CHAPTER 13

1. c	2. a	3. c	4. b	5. a
6. c	7. a	8. c	9. a	10. c

CHAPTER 14

1. a	2. b	3. c	4. d	5. b
6. a	7. d	8. a	9. c	10. d

FINAL EXAM

1. a	2. c	3. a	4. b	5. c
6. d	7. c	8. b	9. a	10. b
11. a	12. a	13. c	14. a	15. c
16. a	17. b	18. d	19. a	20. b
21. d	22. a	23. d	24. c	25. a
26. d	27. c	28. b	29. a	30. c
31. d	32. d	33. c	34. a	35. b
36. a	37. d	38. b	39. a	40. d
41. c	42. d	43. b	44. a	45. b
46. a	47. c	48. a	49. c	50. b
51. a	52. c	53. b	54. a	55. b
56. a	57. b	58. c	59. a	60. a
61. d	62. a	63. b	64. c	56. a
66. b	67. a	68. c	69. b	70. c
71. a	72. c	73. a	74. c	75. b
76. c	77. d	78. a	79. b	80. a

Suggested Additional References

Further Reading

These are some of the book I found useful in writing this volume; you may find them useful, too!

Berg, P. and Singer, M. *Dealing with Genes: The Language of Heredity,* Mill Valley, CA: University Science Books, 1992.

Davies, K. *Cracking the Genome: Inside the Race to Unlock Human DNA,* New York, NY: The Free Press, 2001.

Elrod, S. L. and Stansfield, W. D. *Schaum's Outline of Theory and Problems of Genetics, Fourth Edition,* New York: McGraw-Hill, 2002.

King, R. C. and Stansfield, W. D. *A Dictionary of Genetics,* New York, NY: Oxford University Press, 1997.

Lewin, B. *Genes V,* Oxford: Oxford University Press, 1994.

Yount, L. *Genetics and Genetic Engineering,* New York, NY: Facts on File, 1997.

Web Sites

There is also an enormous amount of information available online. Here are a few excellent sites I've identified. Many more can be found by following links from the ones below.

www.esp.org—Electronic Scholarly Publishing

www.dnalc.org—Gene Almanac

www.kumc.edu/gec/—Genetics Education Center, University of Kansas Medical Center

gslc.genetics.utah.edu—Genetic Science Learning Centre, University of Utah

www.genomenewsnetwork.org—Genome News Network

www.doegenomes.org—Genome Projects of the U.S. Department of Energy Office of Science

www.biology-pages.info—Kimball's Biology Pages (Dr. John W. Kimball)

www.mendel-museum.org—Mendel Museum

www.mendelweb.org—Mendel Web

www.ncbi.nlm.nih.gov—National Center for Biotechnology Information

www.genome.gov—National Human Genome Research Institute

Glossary

Most terms in this book are defined when used for the first time, and many are discussed in great detail thereafter. However, here are short definitions of some of the most important terms you may encounter in this book and in other reference works on genetics, compiled for quick reference.

activator—A protein that enhances the transcription of a gene.

allele—A version of a particular gene.

anti-codon—The sequence of three bases on a transfer RNA molecule complementary to the codon for the amino acid specified by that tRNA.

autosomes—Non-sex chromosomes.

bacteriophage/phage—A virus that infects bacteria.

base—The parts that distinguish one DNA (or RNA) molecule from another. The four bases in DNA are adenine (A), thymine (T), guanine (G), and cytosine (C). Uracil (U) replaces thymine in RNA.

capsid—The protein coat of a virus particle.

carcinogen—A mutagen that can cause cells to become cancerous.

cell—The basic unit of life, capable of growing and multiplying.

chromatin—A substance composed of folded, condensed double-stranded DNA, associated with various special proteins; the substance of which chromosomes are composed.

chromosomes—The structures in cell nuclei where genetic information is coded in the form of highly condensed DNA and associated proteins.

clade—Any defined group whose members share certain characteristics that distinguish them from nonmembers; *i.e.,* a species and its descendants.

clone—Any organism (or in bacteria, colony of organisms) that is/are identical genetic copies of an original.

coding region—A stretch of DNA or RNA that contains a series of codons that can be translated into a polypeptide chain.

codon—Three consecutive nucleotides in DNA or RNA that specify a particular amino acid or indicate the beginning or end of a coding region.

complementary—Bases that can pair with each other are said to be complementary.

Complementary DNA (cDNA)—A DNA chain formed by copying an RNA chain using the enzyme reverse transcriptase

conjugation—The transfer of DNA from one bacterium to another.

cytoplasm—The material inside the cell membrane of a cell, excluding the nucleus.

deoxyribonucleic acid (DNA)—The basis of heredity, a self-replicating molecule in the form of a double helix composed of two sugar-phosphate backbones bound together by pairs of nitrogenous bases.

diploid—A cell that has two copies of each of the chromosomes typical of its species.

DNA fingerprint—The pattern of DNA fragments produced from a specific section of a genome by cleaving it with a particular restriction nuclease then separating the fragments by gel electrophoresis. The pattern may be unique to an individual organism of a species.

DNA polymerase—An enzyme that lengthens a DNA chain by assembling individual nucleotides and joining them via base pairing to another DNA chain, the template.

enzyme—A protein that acts as a catalyst for cell chemistry; it facilitates biochemical reactions but is not itself altered by them.

episome—Genetic elements in bacteria that may either replicate independently of the bacterial chromosome or attach themselves to the chromosome and replicate with it.

equilibrium population—A population in which the frequencies of the alleles in its gene pool don't change through successive generations.

eukaryote—An organism whose cells contain a nucleus.

exon—A portion of a gene that is maintained in functional RNA and that usually contains sequences encoding a polypeptide.

frameshift—A type of mutation caused by the deletion of a base pair, so that the subsequent codons are misread.

gamete—A haploid germ cell.

gene—A segment of DNA or RNA that contains genetic information.

gene expression—The process of conversion of the information contained in a gene into a component of a cell.

genetic code—The language of DNA and RNA, read in three-letter words made up of an alphabet of four bases. Each three-letter codon represents either a specific amino acid or the end of a coding sequence.

genetic map—A visual depiction of the order in which genes occur along a chromosome.

genotype—The genetic makeup of an organism, as opposed to its phenotype (physical characteristics).

germ cells—Cells that combine to give rise to new multicellular organisms, *i.e.,* sperm and egg cells in animals and egg and pollen cells in plants.

haploid—A cell that only has a single copy of each of the chromosomes typical of its species.

heterozygous—The state of having a different allele on each of a pair of chromosomes in a diploid genome.

homologous—Containing the same sequence of either genes (in the case of chromosomes) or base pairs (in the case of DNA).

homozygous—The state of having two copies of the same alleles in each of a pair of chromosomes in a diploid genome.

hormone—A relatively small protein molecule that serves as a signal to coordinate the activities of various cells and tissues in a multicellular organism.

intron—A portion of a gene that does not encode any protein.

library—A population of identical vectors, each of which carries a different DNA which, taken together, may represent an entire genome or specific portion of it.

lysogenic cycle—A viral life cycle that does not kill the host cell but instead alters it to produce viruses on an ongoing basis.

lytic cycle—A viral life cycle that proceeds quickly to destruction of the host cell by lysis (rupturing), which in turn releases many new copies of the virus into the cell's surroundings.

meiosis—The process of chromosome sorting associated with the kind of cell division that leads to new haploid germ cells.

messenger RNA (mRNA)—A single chain of RNA that can be translated into a protein.

metastasis—The spread of cancer cells from one part of the body to another.

mitochondria—An organelle that plays a vital role in cell metabolism but has its own DNA and is capable of reproducing semi-autonomously.

mitosis—The process of chromosome sorting associated with the kind of cell division that leads to new diploid somatic cells.

mutation—Any alteration in DNA structure from one generation to the next.

neoplasm—A group of cells growing out of control, said to be benign if the cells do not metastasize and malignant if they do.

nucleotide—The basic building blocks of DNA and RNA chains.

nucleus—The membraned organelle within a eukaryotic cell that contains the chromosomes.

oncogenes—A faulty gene that can cause a cell to turn cancerous.

organelles—Small inside cells that perform specialized functions for the cell.

ori—The position on a DNA chain where replication begins.

phenotype—The physical expression of an organism's genetic makeup.

plasmid—A ring of double-stranded DNA in a cell that can replicate independently of the cell's genomic DNA.

point mutation—A mutation that alters a single base in a DNA or RNA sequence.

polymerase chain reaction (PCR)—A method of making many copies of a segment of DNA without cloning.

polypeptide—A molecule formed by joining together many (typically 20 or more) amino acids.

probe—A DNA or RNA segment tagged with a radioactive molecule (or some other easily identifiable molecule), used to locate complementary nucleic acid sequences in a library.

prokaryotes—Single-celled organisms that lack a nucleus, *i.e.*, bacteria.

promoter—A DNA sequence that must be present for transcription to begin.

protooncogenes—The normal versions of genes that, in altered form, sometimes cause cancer.

recombinant DNA—A DNA molecule formed via laboratory procedures by assembling DNA segments drawn from various sources, biological or synthetic

repressor—A protein that inhibits transcription of a gene by binding to a specific DNA sequence.

restriction nuclease—An enzyme that splits DNA molecules at a specific sequence.

restriction site—The specific sequence at which a restriction nuclease will split a DNA molecule.

restriction fragment length polymorphisms (RFLPs)—Differences in the size of fragments produced by cleaving the same region of a genome with a restriction nuclease.

ribonucleic acid (RNA)—Similar to DNA, but with the sugar ribose instead of deoxyribose in its backbone and the base uracil instead of thymine.

ribosomes—Particles made of RNA and protein that assemble amino acids into polypeptides, based on RNA templates.

RNA polymerase—An enzyme that lengthens a RNA chain by assembling individual nucleotides as indicated by a complementary strand of DNA, the template.

sex-linked traits—Traits whose genes appear on one of the sex chromosomes but not the other, in species with two different sex chromosomes.

somatic cells—All the cells of the body except for the germ cells.

somatic nuclear transfer—The technique used for cloning higher animals, involving inserting the nucleus from a body cell into an egg cell from which the nucleus has been removed.

start and stop codons—Codons that mark the beginning and end of the DNA sequence that should be translated.

telomere—The end of a chromosome.

transcription—The process by which an RNA polymerase synthesizes an RNA chain complementary to a DNA template.

transduction—The introduction of foreign DNA into a bacterium by a phage.

transfer RNA (tRNA)—Small RNA chains that can each join to a specific amino acid and also have three consecutive nucleotides that are complementary to the codon for that amino acid.

transformation—A permanent and heritable change in the properties of a cell brought about either by the insertion of a new DNA sequence or a mutation.

transgenic—An organism whose genome has been altered by the introduction of new DNA sequences, and which is capable of passing those genetic changes on to its offspring.

translation—The process by which the codons in mRNA are decoded into a polypeptide with a specific sequence of amino acids.

transposable elements—DNA segments that can move from one place to another in a genome; also called transposons, mobile elements, and "jumping genes."

vector—A DNA molecule, usually in a virus or plasmid, that can be joined with a foreign segment of DNA so that the new DNA can be introduced into a cell.

virion—A complete virus, with protein coat and genetic material.

wild-type—The most frequently observed genotype or phenotype, *i.e.,* "normal."

zygote—The diploid cell produced by the fusion of the male and female haploid gametes.

INDEX

ABOUT THE AUTHOR

Edward Willett is a science columnist for radio and newspapers and a former news editor. The author of more than 30 books, including nonfiction on topics as diverse as computing, disease, history, and quantum physics, as well as several science fiction and fantasy novels, he is the recipient of awards from the National Science Teachers Association, the Children's Book Council, and *VOYA* magazine, among others.